Primate Retina and Choroid

Wolf Krebs
Ingeborg Krebs

Primate Retina and Choroid
Atlas of Fine Structure in Man and Monkey

With 65 Illustrations

Springer-Verlag

Wolf Krebs, DVM adj., Associate Professor
and
Ingeborg Krebs, MD, Research Scientist
Department of Cell Biology and Anatomical Sciences
The City University of New York Medical School
New York, NY 10031, USA

Frontcover: Junctional complex of a foveal pigment epithelial cell in tangential section.

Library of Congress Cataloging-in-Publication Data
Krebs, Wolf.
Primate retina and choroid: atlas of fine structure in man and monkey / Wolf Krebs, Ingeborg Krebs.
p. cm.
Includes bibliographical references.
Includes index.
ISBN 0-387-97432-6 (alk. paper). — ISBN 3-540-97432-6 (alk. paper)
1. Retina—Ultrastructure—Atlases. 2. Choroid—Ultrastructure—Atlases. 3. Monkeys—Anatomy—Atlases. I. Krebs, Ingeborg. II. Title.
[DNLM 1. Electron microscopy—Atlases. 2. Choroid—ultrastructure. 3. Primates—anatomy & histology. 4. Retina—ultrastructure. WW 17 K929p]
QM511.K44 1991
599.8'0448—dc20
DNLM/DLC
for Library of Congress 90-10376

Printed on acid-free paper.

Softcover reprint of the hardcover 1st edition 1991

Typeset by Best-set Typesetter, Ltd., Hong Kong.

9 8 7 6 5 4 3 2 1

ISBN-13: 978-1-4613-9097-8 e-ISBN-13: 978-1-4613-9095-4
DOI: 10.1007/978-1-4613-9095-4

Preface

The electron micrographs selected for this atlas have been taken during 5 years of research on the primate retina in the Department of Ophthalmology of Columbia University. This project would not have been possible without financial support by Manhattan Eye, Ear, and Throat Hospital, New York, and it would not have been completed without Basil Worgul who sided with the authors whenever their work was endangered. Also helpful have been Raul Chiesa, Shonit Das, Anthony Donn, Peter Gouras, Larry Liebowitch, and John Merriam. The authors wish to thank all of these and many other unnamed colleagues and friends who have been supportive and encouraging. Thanks are due to the editorial and production staff of Springer-Verlag. They not only agreed to publish this book, but their advice also significantly influenced its contents. The permission to use previously published material was kindly given by the publishers of the American Journal of Anatomy (Table 1), Archives of Ophthalmology (Fig. 59), and Histology and Histopathology (Fig. 49).

Contents

Introduction

This volume describes the morphology of the primate retina as seen with the electron microscope. As it is an atlas, the electron micrographs are its most important part. The text accompanies the figures, highlighting selected topics either to explain structures or to point out structure-function relationships. A scholarly review of the whole spectrum of research on the retina and choroid is not feasible in a single volume. Thus, whenever available, review articles or monographs, rather than original work, are cited for reference.

The introductory chapters present some basic information on ocular morphology and on the embryology and histology of retina and choroid. They also define the morphological terms used in this book. The proper anatomical names, as listed in *Nomina Anatomica* (1983), are given in brackets at the first mention of the corresponding structure. In the following text, the English translation of the anatomical name is used. An exception is "Bruch's membrane," which properly should be called "basilar complex." However, this anatomical name is so rarely used in the ophthalmic literature that "Bruch's membrane" is preferred.

The readers who want to refresh their knowledge of histology and fine structure of mammalian tissue may consult any modern textbook of histology and microscopic anatomy. The books by Fawcett (1966), Porter and Bonneville (1973), and Rhodin (1974) are recommended in particular. Histology of the Human Eye by Hogan, Alvarado, and Wedell (1972) not only describes the tissues of the eye, including retina and choroid, it also contains a chapter dealing with the ultrastructure of cells. Another excellent survey of normal cellular and extracellular ultrastructure can be found in Ultrastructural Pathology of the Cell and Matrix by Ghadially (1982).

The Primate Eye

The principal part of the visual organ [*organum visus*] is the eyeball [*bulbus oculi*], also called bulb or globe (**Fig. 1**). It is suspended within the orbital cavity of the skull by six ocular muscles. The wall of the orbit, the eyelids, the conjunctiva, and the lacrimal apparatus are the protective structures of the eye.

Three tunics make up the wall of the eye ball. The innermost tunic [*tunica interna bulbi*] is the retina (solid black area in Fig. 1). Anteriorly it begins at the rim of the pupil (Fig. 1: arrow). It covers the posterior part of the iris (Fig. 1: I) as the bilayered iris epithelium [*pars iridica retinae*]. The retina then spreads over the ciliary body [*corpus ciliare*] (Fig. 1: CB), still bilayered, as the ciliary epithelium [*pars ciliare retinae*]. At the *ora serrata* (Fig. 1: OS) the retina abruptly becomes thicker.

The *ora serrata* marks the border between two parts of the retina: (1) the anterior *pars caeca* (Fig. 1: PC), the bilayered iris and ciliary epithelium, and (2) the posterior *pars optica retinae* (Fig. 1: PO), the seeing part of the retina. The optical retina covers the posterior four-fifths of the internal surface of the wall of the globe. It is perforated by the passage of the optic nerve through the wall of the globe near its posterior pole. This interruption of the retina is visible as the papilla of the optic nerve (Fig. 1: P), also known as blind spot or optic disc.

The middle tunic of the wall of the eyeball is called the *uvea* [*tunica vasculosa bulbi*]. It too is limited by the pupil anteriorly and traversed by the optic nerve posteriorly. The uvea has three parts: the iris (Fig. 1: I), ciliary body (Fig. 1: CB), and choroid [*choroidea*]. The choroid is the posterior part of the uvea that is in contact with the optical retina.

The third tunic [*tunica fibrosa bulbi*] forms the corneal stroma and sclera. The fibrous tunic, composed of tightly interwoven collagen fibers, provides protection to the tissues of the eyeball and support for its adnexes.

In this volume, the following topographical terms are used: The anterior pole of the eyeball is the center of the cornea, which then defines the posterior pole at the opposite side of the eyeball. The equator of the globe is at the intersection of the bulbar surface with a plane through the center of the globe, perpendicular to a line connecting anterior and posterior poles. An equatorial section is parallel to that plane. A radial section is a section along a plane through the center of the eyeball. A tangential section is defined as a section parallel to a plane tangential to the surface of the globe. The terms external (or outer) and internal (or inner) describe positions that are closer to the exterior surface or to the center of the eyeball than to a reference structure.

Figure 1. Diagram of primate eye. CB, ciliary body; F, fovea centralis; I, iris; OS, ora serrata; P, *papilla nervi optici*; PC, *pars caeca retinae*; PE, pigment epithelium; PO, pars optica retinae; arrow, rim of pupil.

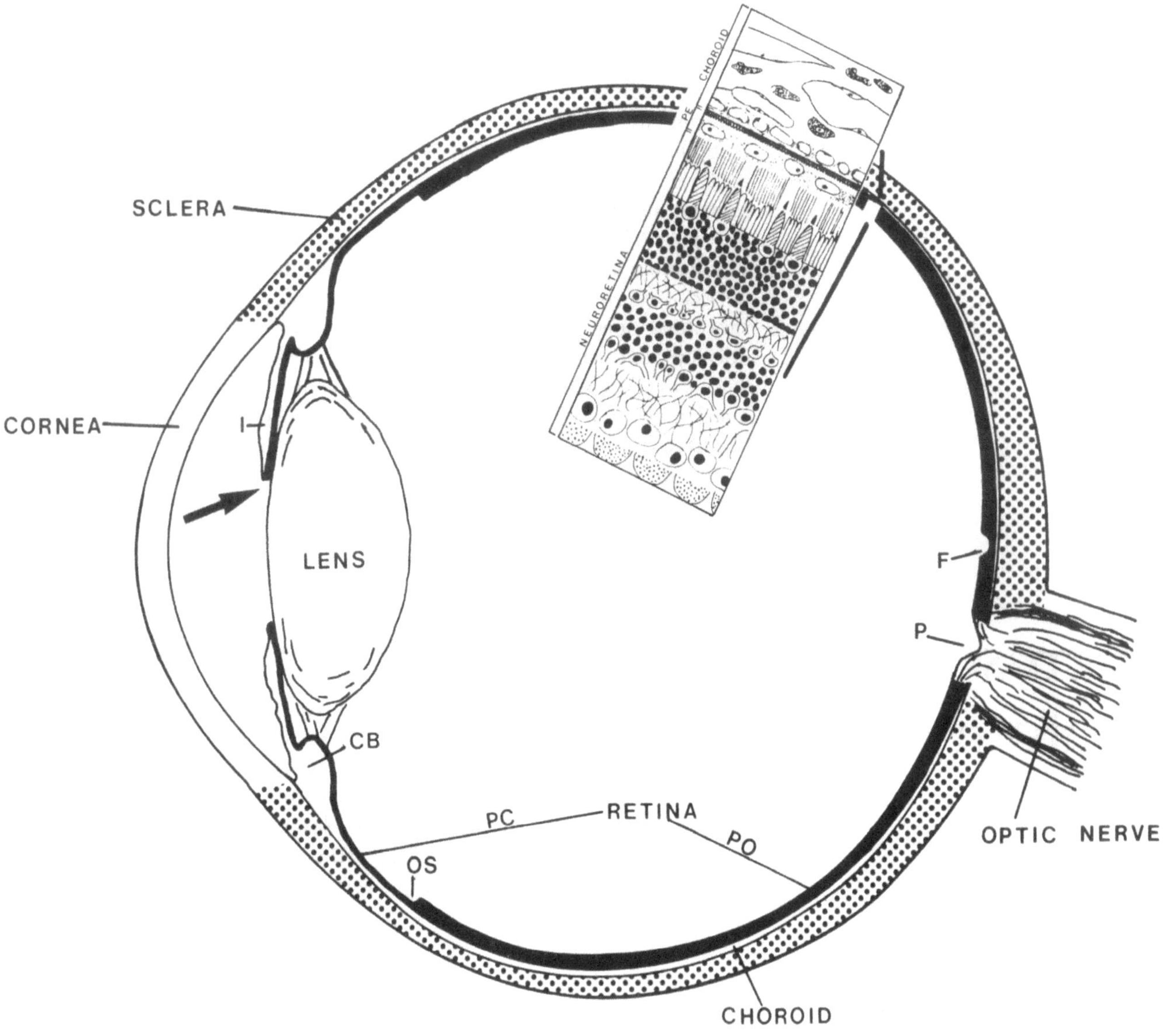

CHOROID
PE
NEURORETINA
SCLERA
CORNEA
I
LENS
F
P
CB
PC
RETINA
PO
OS
OPTIC NERVE
CHOROID

Embryology of Retina and Choroid

The retina is a part of the brain. At the end of the first month in the development of the human embryo, two optic vesicles develop at the lateral surfaces of the embryonic forebrain. The lumina of the optical vesicles are continuous with the primitive ventricular system. At the end of the fourth week, the optical vesicles collapse and form optic cups. The ventricular space is reduced to a capillary cleft between two sheaths of neuroectodermal epithelium. The optical cups, connected to the forebrain by the optical stalks, induce the formation of the eyes in a complicated sequence of events (see for example: Starck 1975). In the adult, the anterior part of the retina, covering the ciliary body and iris, retains the original structure of two epithelial sheaths; the apical poles of their cells face each other. In the posterior part of the retina, the outer neuroectodermal sheath of the eyecup remains a single layer. It becomes the retinal pigment epithelium, whereas the inner sheath transforms into the nervous tissue of the neuroretina. The former optical ventricle remains a distinct extracellular compartment between the neuroretina and pigment epithelium: It is commonly termed subretinal space. A summary of recent research on vertebrate retinal development can be found in Barnstable (1987).

The uvea develops from the mesenchyme surrounding the eyecup. The uveal melanocytes originate from the neural crest. They immigrate into the uvea by the fourth embryonic week in humans. Thus, retinal pigment epithelial cells and uveal melanocytes belong to different cell lines. As the retina is part of the brain, the uvea is homologous to the pia mater (Shanthaveerappa and Bourne 1965). Details of the development of the choroid can be found in Heimann (1972) (human) and in Ozanics et al. (1978) (macaque).

Microscopic Anatomy

Retina

As we will consider only the *pars optica retinae*, we will refer to it from now on simply as the retina. Microscopically, in radial sections, the retina appears stratified (**Fig. 2**). The outermost stratum is the pigment epithelium [*stratum pigmentosum*] (Fig. 2: PE). It consists of a monolayer of cuboidal pigmented cells. The remaining nine strata are collectively called neuroretina [*stratum nervosum*]. The outer layer of cells in the neuroretina are the photoreceptor cells [*epitheliocyti neurosensorii*]. They are slender cells oriented normally to the plane of the retina. There are two kinds: rod cells [*epitheliocyti bacilliferi*] (Fig. 2: RC) and cone cells [*e. coniferi*] (Fig. 2: C).

Figure 2. Schematic representation of retina and choroid. (A) As seen in histological section. (B) Retinal neurons shown diagrammatically. A, amacrine cells; B, bipolar cells; BM, Bruch's membrane; C, cone cells; CC, choroidocapillaris; ELM, external limiting membrane; G, ganglion cells; GCL, ganglion cell layer; H, horizontal cells; ILM, inner limiting membrane; INL, internal nuclear layer; IS, inner segments of rods and cones; IPL, internal plexiform layer; NFL, nerve fiber layer; ONL, outer nuclear layer; OPL, outer plexiform layer; OS, outer segments of rods and cones; PE, pigment epithelium; PRL, photoreceptor layer; PT, photoreceptor cell terminals; RC, rod cells; ST, *stroma vascularis* of choroid.

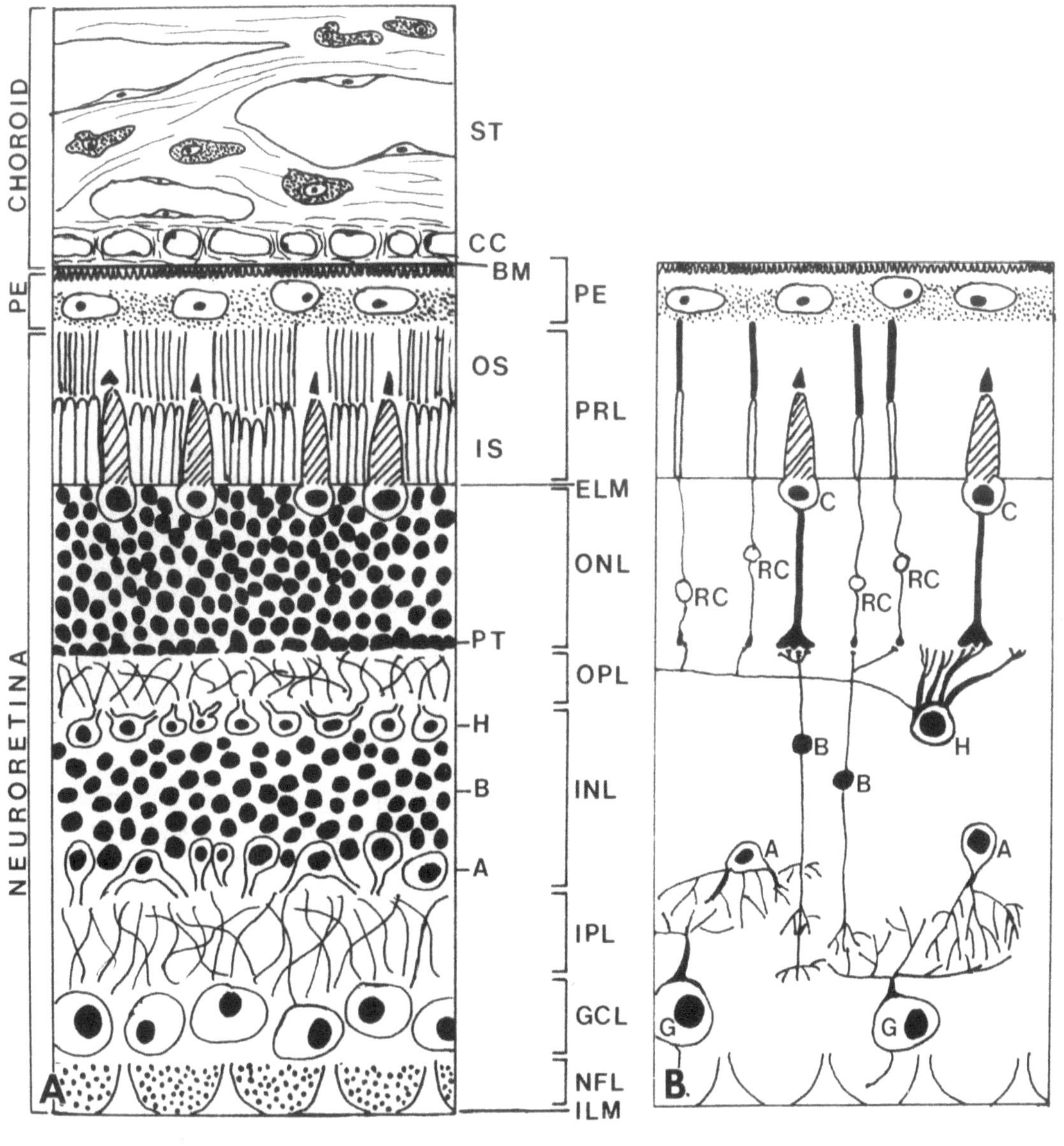
CHOROID
ST
CC
BM
PE
PE
OS
PRL
IS
ELM
ONL
PT
NEURORETINA
OPL
H
INL
B
A
IPL
GCL
NFL
ILM
A
B.
C
C
RC
RC
RC
RC
H
B
B
A
A
G
G

Rod and cone cells share a principal design, which consists of an apical light-sensitive part, the rod [*bacillus*] and the cone [*conus*], respectively. Rods and cones are arranged in parallel to make up a distinct stratum [*stratum photosensorium*] (Fig. 2: PRL). Both rods and cones are subdivided into an outer segment [*segmentum externum*] (Fig. 2: OS), a connecting cilium, and an inner segment [*segmentum internum*] (Fig. 2: IS). The outer segment contains tightly stacked, flattened membranous vesicles [*disci membranacei*], stacked in register, which contain the visual pigment rhodopsin. Morphologically, rods differ from cones in that they are more slender and longer; as the names imply, cones are conically shaped and rods are rod-like.

At the bases of rods and cones, the photoreceptor cells are attached to the apices of adjacent retinal radial glial cells [*gliocyti radialis*] by *zonulae adherentes* that are continuous with adhering junctions between the radial glial cells. The radial glial cells commonly are called Müller cells. The system of adhering junctions is visible in radial light microscope sections as a stippled line that is regarded as the third stratum of the retina [*stratum limitans externum*] (Fig. 2: ELM). Schultze (1866) is credited with the discovery of this structure and he coined the word *membrana limitans externa*, a name still used most frequently in its English translation: outer or external limiting membrane. The retinal radial glial cells (Müller cells) extend microvilli into the subretinal space.

Internally from the outer limiting membrane is the stratum of photoreceptor cell perikarya [*stratum nucleare externum*] (Fig. 2: ONL), which in English usually is called the outer or external nuclear layer. The nuclei of the photoreceptor cells are the most prominent constituents of this stratum, which also contains the following: the outer fibers of the cone and rod cells, which connect the perikarya to the rods and cones; the photoreceptor cell inner fibers, which lead to the synaptic terminals; and cytoplasmic extensions of the radial glia, tightly filling the spaces between the photoreceptor cells.

The synaptic terminals of the rod and of the cone cell are the rod cell spherule [*spherula terminalis*] and the cone cell pedicle [*pes terminalis*] (Fig. 2: PT). Collectively they mark the border to the next stratum, the outer or external plexiform layer [*stratum plexiforme externum*] (Fig. 2: OPL).

Some authors (e.g., Polyak 1941, Fig. 36) include the photoreceptor terminals in the outer plexiform layer; others (e.g., Dowling 1987, p. 14, these authors) regard them as part of the outer nuclear layer. In the external plexiform stratum, cellular processes of the second set of retinal neurons, horizontal and bipolar cells [*neurona horizontalia et bipolaria*] (Fig. 2: B,H), intermingle and contact the photoreceptor cell terminals.

The cell bodies of the horizontal cells are located in the external part of the inner nuclear layer [*stratum nucleare internum*] (Fig. 2: INL). Primate horizontal cells have axons which, as their dendrites, enter the external plexiform layer (Fig. 2B:H).

In the central part of the inner nuclear layer are the perikarya of the radial glial cells (Müller cells) and the cell bodies of bipolar cells. The dendrites of bipolar cells course toward the external plexiform layer in which they arborize (Fig. 2 B: B). Their axons pass along the cell bodies of the inner nuclear layer toward the next stratum, the internal plexiform layer [*stratum plexiforme internum*] (Fig. 2: IPL).

The perikarya of amacrine cells [*neurona amacrina*] (Fig. 2: A) form a sublayer at the inner part of the inner nuclear layer. Amacrine cells lack axons. Their dendrites penetrate into the inner plexiform layer.

There is an additional neuron that received considerable attention under the name "interplexiform cell" (Dowling 1987, pp. 29 ff). Its cell body is located between the perikarya of amacrine cells. It was discovered by Cajal (1904) and extensively discussed by Polyak (1941, pp. 298 ff) under the name "centrifugal bipolar cell." The dendrites of this cell come out of the inner plexiform layer, and its axon extends toward the external plexiform layer.

In the inner plexiform layer, the dendrites of retinal ganglion cells [*neuronum multipolare*] (Fig. 2B: G) intermingle with axons of bipolar cells and the processes of amacrine cells. The inner plexiform layer is thicker than the outer plexiform layer.

In the ganglion cell layer [*stratum ganglionare*] (Fig. 2: GCL), the perikarya of retinal ganglion cells predominate. The axons of these cells cover the inner surface of the retina [*stratum neurofibrarum*] (Fig. 2: NFL), separated from the vitreous by the basilar membrane [*stratum limitans internum*] of the radial glia (Fig. 2: ILM).

Primates are the only mammalian species with a *fovea centralis* of the retina (**Fig. 3**). The fovea is at the intersection of retina and optical axis of the globe (Fig. 1: F). The fovea is the area of the most acute vision. The thickness of the retina is reduced in the fovea because the photoreceptor cell synapses and the inner retinal neurons are displaced peripherally from the foveal center. Cones are concentrated in the fovea; its center is free of rods. Peripherally the number of rods increases, reaching a maximum at the perimeter of the fovea (Østerberg 1935). The diameter of the fovea in humans is 1.5 mm; and the central part, which is free of the inner layers of the retina, is 0.35 mm across (Hogan et al. 1972, pp. 49 ff) and is called foveola. The tissue between the foveola and foveal rim, wedge-like in cross section, is called the *clivus* or foveal slope. The axons or outer fibers of the foveal rod and cone cells are elongated and form an additional layer, the "fiber layer of Henle," between the outer nuclear and outer plexiform layer in the periphery of the foveal area (Fig. 3: HFL). The foveal photoreceptor cells, including the fibers of Henle, contain the yellow pigments zeaxanthin and lutein (Bone et al. 1988). Thus, the area surrounding and including the fovea has a distinct hue. It is called *macula lutea*, the yellow spot. Often "macula" and "fovea" are wrongly used as synonyms.

Blood is supplied to the neuroretina, photoreceptor cells excepted, via retinal vessels. The large arteries and veins lie within the nerve fiber layer. Ascending arteries penetrate into the retinal tissue. Two flat beds of capillaries spread between the perikarya of horizontal and of amacrine cells, at the outer and inner margins of the internal nuclear layer. Another network of capillaries supplies the ganglion cell and the nerve fiber layer.

Choroid

The choroid provides the blood supply for the pigment epithelium and photoreceptor cells of the retina. Somewhat arbitrarily, it can be subdivided into layers. The innermost layer commonly is called Bruch's membrane [*Complexus basalis*] (Fig. 2A: BM). It consists of the basilar lamina of the retinal pigment epithelium, an inner and an outer layer of collagen fibers [*stratum fibrosum*], and a central layer of elastic material [*stratum elasticum*]. The next layer of the choroid is a sinusoidal system of blood capillaries, the choroidocapillaris [*lamina choroidocapillaris*] (Fig. 2A: CC). In the ophthalmic literature this structure frequently is termed the choriocapillaris. However, in anatomical nomenclature, the prefix "chorio" is used for structures of the chorion.

Externally of the choroidocapillaris is the *substantia propria* of the choroid, which also is termed the vascular stroma (Fig. 2A: ST). As the latter name implies, the main constituent of this part of the choroid is blood vessels, namely veins and arteries but no capillaries. The vascular stroma is rich in nerves. It contains the choroidal melanocytes [*melanocyti choroidei*] and the typical cells of loose connective tissue. The occurrence of ganglion cells in the stroma of the choroid has been reported (Kurus 1955). The outermost layer of the choroid is a thin sheath of flattened cells, the suprachoroid [*lamina suprachoroidea*]. A perichoroidal cleft [*spatium perichoroidale*] separates the choroid from the sclera. A survey of the choroidal literature before 1980 can be found in Torczynski's (1982) review article.

Figure 3. Schematic representation of fovea. GCL, ganglion cell layer; HFL, Henle's fiber layer; INL, internal nuclear layer; ONL, outer nuclear layer, rod cell nuclei black; OPL, outer plexiform layer; PRL, photoreceptor layer.

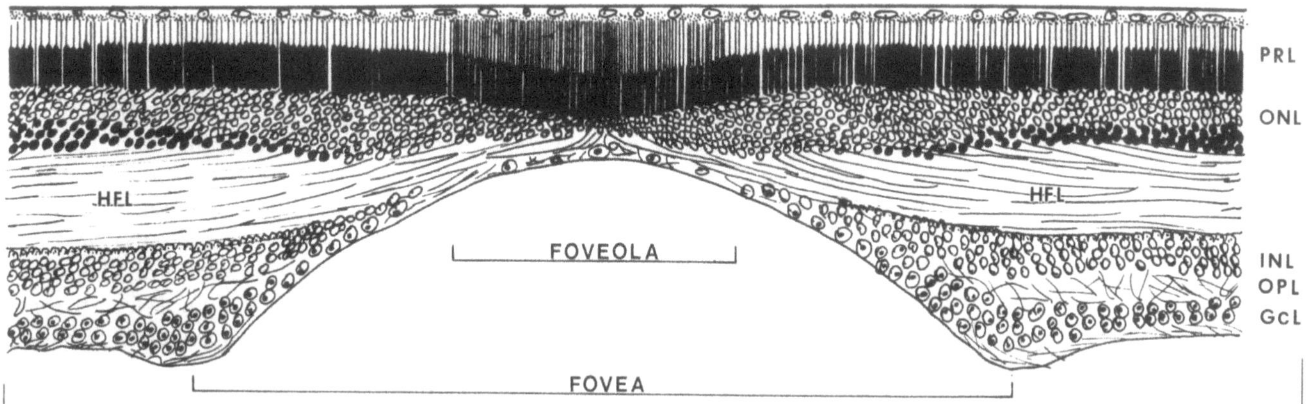
PRL
ONL
HFL
HFL
FOVEOLA
INL
OPL
GcL
FOVEA

Material and Methods

One human eye and the eyes of five cynomolgus monkeys (*Macaca irus*), one baboon (*Papio anubis*), and one pygmy marmoset (*Cebuella pygmea*) are the sources of the tissues used for this book. The human eye was supplied from the Eye Bank for Sight Restoration, New York. It was enucleated 3 hours after death from a 58-year-old white man who suffered from Hodgkin's disease. The 1/2-year-old pygmy marmoset came from the Bronx Zoological Garden, New York. Its eyes were extirpated immediately after death. They were fixed in Trump's solution in the Pathology Laboratory of the New York Zoological Society. The cynomolgus and baboon eyes were obtained from animals used for experimental heart surgery at the Columbia Presbyterian Medical Center. These eyes were enucleated from the narcotized animals at the end of the experiments and before euthanasia.

As published (Krebs and Krebs 1987, 1989a), the eyes were cut open at the *limbus corneae* and, with the exception of the pygmy marmoset eye, immersed in a fixative of 2% glutaraldehyde and 2% formaldehyde in 150 mM cacodylate buffer at pH 7.4 containing 10 mM magnesium chloride. After a minimum of 12 hours fixation and several rinses with buffer, the anterior segments of the eyes were removed.

A 1 mm trephine was used to punch out pieces of retina. These retinal plugs contained retina and choroid. Samples were taken from defined retinal loci as depicted in **Figure 4**. One contained the fovea, four were taken at a distance of 1 mm from the rim of the foveal sample, and four or five were punched out at the nasal segment of the retina at the same distance from the optic nerve as the foveal sample. These loci will be designated as foveal, parafoveal, and peripheral retina. The peripheral sample was located at 30 degrees nasal eccentricity.

Figure 4. Drawing of background of human eye. Positions from which samples were taken are indicated by black circles. F, foveal sample; OD, optic disc; P, peripheral sample at 30 degrees nasal eccentricity; PF, parafoveal sample.

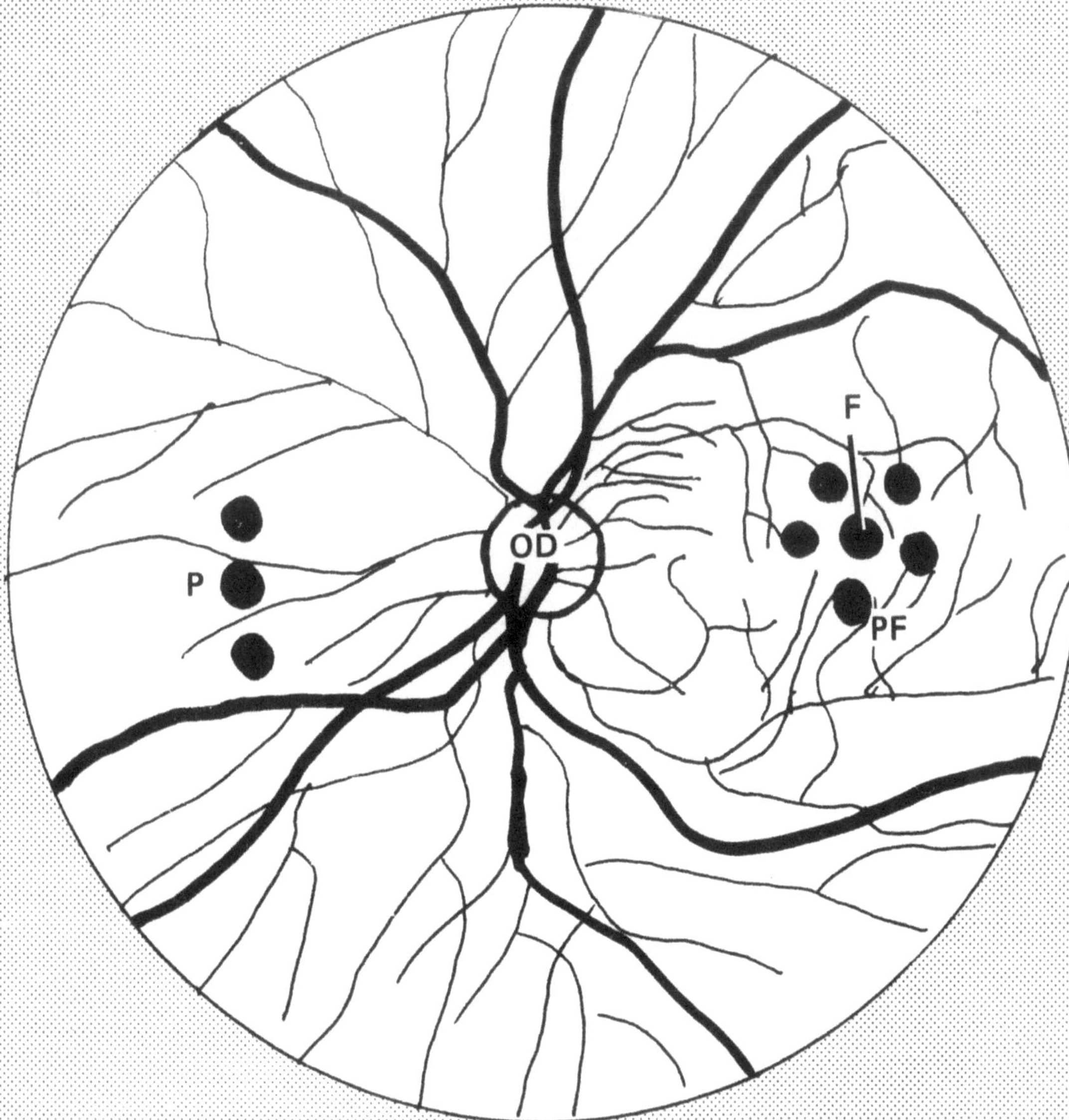
F
OD
P
PF

The specimens were stained (postfixed) with osmium tetroxide, dehydrated, and embedded in epoxy resin. There was virtually no shrinkage in the diameter of the retinal plugs. **Figure 5** shows an embedded plug together with the tip of the trephine used to punch out the tissue sample. Radial and tangential sections were cut from the embedded tissue. Semithin (1 μm) sections were stained with toluidine blue and basic fuchsin. Ultrathin (0.1 μm) sections were stained with uranyl acetate and lead citrate.

Morphometric measurements were made in montages of electron micrographs of tangential sections taken from the appropriate levels within the retina as indicated in Figure 6. A semiautomatic, computerized image analysis system was used. Most cellular densities were determined at magnifications of 3,000 X. Bipolar cells were counted at 6,000 X magnification. For further details consult Krebs and Krebs 1987. The figures describing the foveal cone cell pedicles, foveal horizontal, and foveal bipolar cells were taken in sections of the appropriate level 500 μm from the foveolar center (see Fig. 7: cross).

Figure 5. Retinal sample embedded in block of epoxy resin together with tip of 1mm trephine used to punch out sample. R, piece of retina and choroid, embedded in cylindrical epoxy block; T, tip of trephine, 1mm diameter.

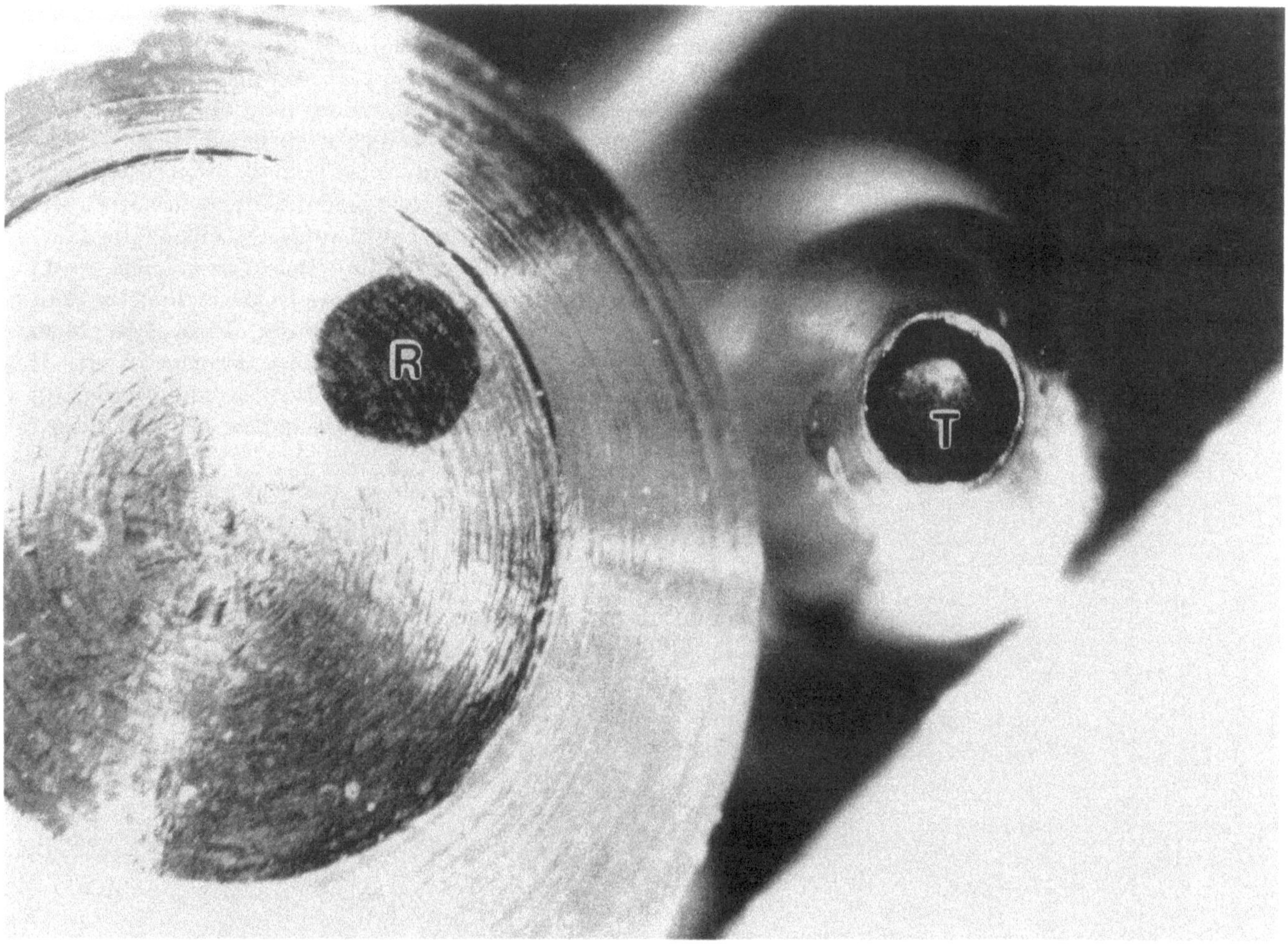
R
T

Fine Structure of the Retina

Figure 6 depicts radial light microscopic sections of the baboon retina at the regions examined. The area of the foveal rim is shown in Figure 6A. At the arrowhead at the bottom is the thickest part (455 μm) of the retina. The shape of the fovea can vary considerably among individuals (Dawson et al. 1986). In this specimen, which has a very steep foveal slope, the foveal center is approximately 0.1 mm toward the left from the arrowhead (Fig. 6A). The downward-pointing arrow at the right indicates the retinal level internal of which the tangential sections are cut to study the inner retinal layers connected to the fovea by Henle's fibers (h). The tissue columns used for this purpose are taken from a region approximately 500 μm from the center. At the foveal rim the thickness of the inner nuclear layer is 114 μm. Approximately 200 μm from the center, it tapers down to 75 μm. This thickness does not change for the next 1 mm. The ganglion cell layer (g) is the thickest just in the region, from which the tangential section of the foveal inner layers are taken.

Figure 6B shows the parafoveal sample taken at a distance of approximately 1 mm from the region seen in Figure 6A. In this region the retina is 340 μm thick. Most of the reduction in thickness is in Henle's fiber layer (h), which is reduced from 95 to 38 μm; and in the ganglion cell layer (g) (95 versus 38 μm). In the peripheral sample (Figure 6C), the retina is merely 208 μm thick. There is no Henle's fiber layer and all retinal layers are thinner than those in the parafoveal and foveal samples. The numbers at the right border of Figure 6C refer to the numbers of the Figures that follow, depicting tangential sections taken at the levels indicated by the arrows. The choroid varies in thickness from 95 μm in the fovea to 72 μm in the parafovea and 55 μm in the periphery.

Figure 6. Light micrographs of radial sections through retina of baboon, representing loci from which retinal samples were taken, ×264. (A) Fovea. (B) Parafovea. (C) Periphery (30 degrees nasal eccentricity). Numbers at right border, figure numbers of tangential sections representing indicated levels within retina; C, choroid; e, external limiting membrane; en, external nuclear layer; g, ganglion cell layer; h, Henle's fiber layer; i, internal limiting membrane; in, internal nuclear layer; ip, internal plexiform layer; is, rod and cone inner segments; n, nerve fiber layer; os, rod and cone outer segments; p, cone cell pedicles; pe, pigment epithelium; downward arrow, region in which foveal cone cell pedicles and foveal inner retinal cells have been studied; arrowhead, rim of fovea.

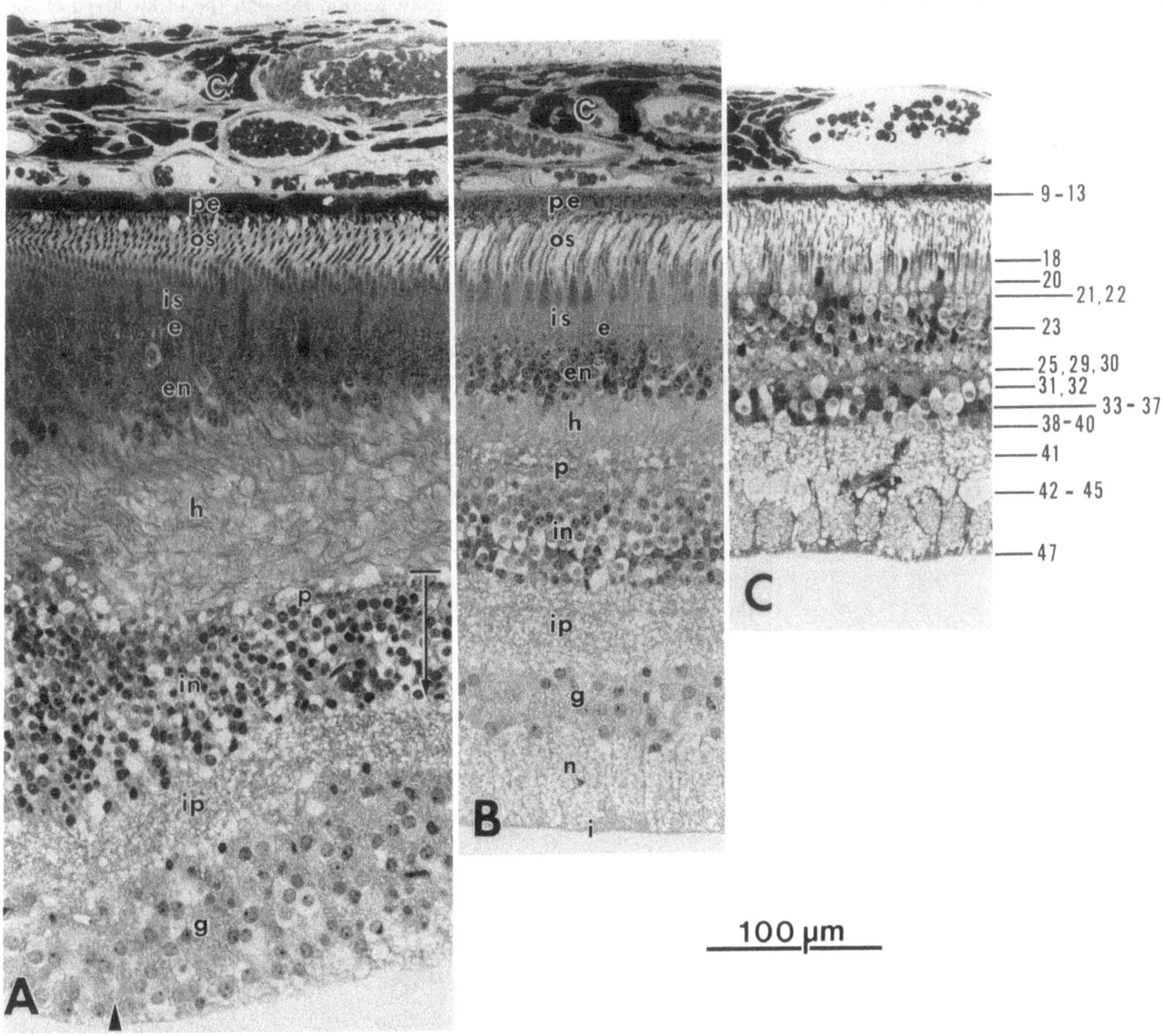

C
pe
os
is
e
en
h
p
in
ip
g
A
B
n
i
9 - 13
18
20
21, 22
23
25, 29, 30
31, 32
33 - 37
38-40
41
42 - 45
47
100 µm

Figure 7 is a tangential semithin section through the other fovea of the baboon. In the center of the fovea (C), the section cuts through the innermost perikarya of the central cone cells. Their axons, known as Henle's fibers (h), radiate into the periphery. They terminate with cone cell pedicles (P), which are arranged in a regular monolayer. Adjacent to the pedicles, horizontal cells (H) are in the plane of the section. At this level the flat network of blood capillaries (CA) is seen. Further to the periphery and deeper in the retina, bipolar cell nuclei (B) are seen. The cross marks a distance of 500 μm from the center. Tangential sections of foveal cone cell pedicles and of the inner retinal cells, which are connected to them, were taken at this position.

Retinal Pigment Epithelium

Considerable research efforts are focused on the retinal pigment epithelium (Zinn and Marmor 1979). This epithelium is indispensable for the function of retinal photoreceptor cells. It also is relatively accessible to experimentation.

The retinal pigment epithelium is the morphological substrate of the blood-retina barrier between choroid and retina. Thus, it contributes to the maintenance of retinal homeostasis. The pigment epithelium separates two distinct ionic compartments, the interior of the eyeball with retina and vitreous, from the choroid and sclera. The distribution of ions in these compartments generates a voltage between the anterior and posterior pole of the eye, the so-called standing potential (Steinberg 1987). The activity of the

Figure 7. Light micrograph of tangential section through fovea of baboon, ×350. B, bipolar cell nuclei; C, center of foveola; CA, blood capillaries; H, horizontal cell perikarya; h, Henle's fibers; P, cone cell pedicles; cross, distance of 500 μm from foveolar center.

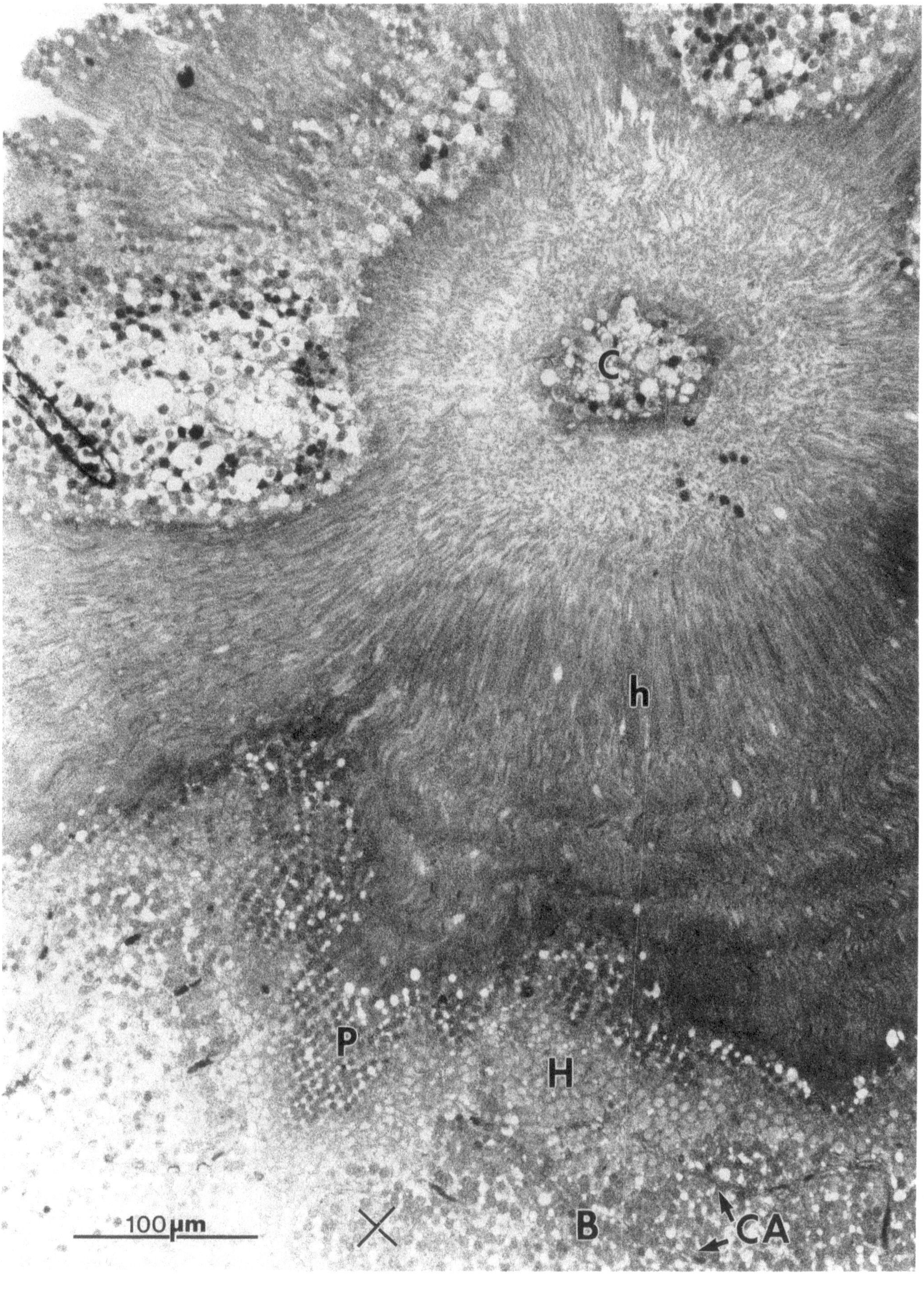
C
h
P
H
100 µm
B
CA

retinal cells generates fluctuations in this voltage, which can be recorded as an electroretinogram (ERG) (Marmor and Lurie 1979). Electroretinography is an important diagnostic tool to assess damage to the retina.

As the retinal pigment epithelium is between choroidal blood supply and neuroretina, it controls the exchange of nutrients and metabolites between the blood and the photoreceptor cells. A well studied example is the vitamin A metabolism of the retina.

Vitamin A (retinoids) in the form of eleven-*cis* retinal is the key molecule in photoreception. It is the prosthetic group of the visual pigment rhodopsin in the outer segments of rods and cones. The biochemical machinery of the vitamin A metabolism resides in the pigment epithelial cells. After light absorption, 11-*cis* retinal isomerizes into its all-*trans* isomer. In this form, it separates from the rhodopsin molecule and is reduced to its alcohol form, retinol. It is then transported into the pigment epithelium cell and reisomerized to the 11-*cis* isomer.

All-*trans* retinol from the blood is taken up also and isomerized and oxidized into 11-*cis* retinal by pigment epithelial cells. Because retinoids are lipophilic, various transport proteins are needed to bridge aqueous compartments inside and outside the cells. The transport of retinoids from blood to photoreceptor outer segments has been intensively studied (for review see Bok 1985). A study with cultured human retinal pigment epithelium presented evidence that the transport of 11-*cis* retinal from pigment epithelium to photoreceptors is mediated by protein (Das et al. 1990).

Pigment epithelial cells are also involved in the metabolism of rod and cone outer segments. They phagocytize the tips of these organelles in synchrony with their growth. Conditions in which the pigment epithelium is unable to phagocytize, lead to destruction of the photoreceptor cells. This abnormality causes blindness in some forms of the disease *retinitis pigmentosa* (Bok 1985).

Macromolecules of the subretinal space, synthesized in the pigment epithelial cell include mucopolysaccharides, predominantly chondroitin sulfate. These compounds have been equated with hypothetical "sticky substances" holding neuroretina and pigment epithelium together (Zauberman 1979). Another prominent molecule of the interphotoreceptor space is the interphotoreceptor retinol binding protein, usually referred to by the acronym IRBP (Bok 1985). As its name implies, this molecule is involved in retinol transport.

The many activities of the pigment epithelium require energy. Thus, it is not surprising that the oxygen consumption of the pigment epithelium was found to be so high in the cat that the oxygen tension falls from 120 mm Hg in the choroidocapillaris to 5 mm Hg or less at the level of rod and cone inner segments (Steinberg 1987).

Figure 8 shows electron micrographs of radial sections through the retinal pigment epithelium. A baboon retinal pigment epithelial cell from a peripheral sample is seen in Figure 8A. The basal border is covered by a basilar membrane (B), which is indented by two small "drusen" (D). Drusen occur in older individuals. They lie between basilar membrane and the collagen layer of Bruch's membrane (BM) and are believed to represent accumulations of residual material. They can become quite large (Hogan et al. 1972, pp. 356 ff).

The system of basal infoldings of the plasmalemma is called the basal labyrinth (BL). It is rather inconspicuous in radial sections of primate pigment epithelial cells, as compared to other species (Kuwabara 1975). Mitochondria (M) are numerous in pigment epithelial cells. They are concentrated in the basal portion of the cells. Lipofuscin granules (LG) or residual bodies accumulate with age in almost any cell type (Ghadially 1982, pp. 452 ff). They are the remnants of lysosomes, containing incompletely degraded material. In the pigment epithelial cells lipofuscin granules most likely are the remnants of phagosomes (PS) containing photoreceptor outer segment membranes. Thus, lamellar inclusions (inset) can be recognized frequently within these granules. Lipofuscin granules are usually seen in the basal part of the pigment epithelial cells.

The nucleus (N) is globular and centrally located. Binucleated retinal pigment epithelial cells were found in all specimens presented here. Ts'o and Friedman (1967) have estimated that 4% of the pigment epithelial cells in the human have two nuclei.

Pigment granules in primates are concentrated in the apical portion of the cells. There are two types of pigment granules: round ones (P_1), forming a broad band in the apical third of the cells, and elongated granules (P_2), located in apical cytoplasmic extensions. In contrast to most other vertebrates, mammals do not exhibit pigment migration (Burnside and Laties 1975).

The cells of the pigment epithelium are held together by an elaborate junctional complex, shown at higher magnification in Figure 8B. The complex consists of a broad *zonula adherens* (Z) and a *zonula occludens* (T). In addition, gap junctions and desmosomes have been reported in primate pigment epithelium (Kuwabara 1975).

Solitary cilia can be seen in many pigment epithelial cells (Fig. 8C). Located in pouches of the apical surface, they have a well-developed basal body (BB) and a striated rootlet fiber (R).

Most prominent in the cytoplasm of primate pigment epithelial cells is the smooth endoplasmic reticulum (SER). It occupies almost any space between the other cellular organelles and consists of tubules too small to be resolved at the magnification of Figure 8A (×9,700). Rough endoplasmic reticulum (RER) and free ribosomes (FR) are found between the pigment granules in the apical cytoplasm.

Figure 8. Radial sections of retinal pigment epithelium. (A) Pigment epithelial cell of peripheral retina of baboon, ×9,700; inset: Lipofuscin granule, ×20,800. (B) Junctional complex, parafovea, baboon, ×51,250. (C) Cilium at apical pole of pigment epithelial cell in peripheral retina of cynomolgus monkey, ×52,500. AM, apical microvilli; B, basilar membrane; BB, basal body; BL, basal labyrinth; BM, Bruch's membrane (*complexus basalis*); D, drusen; FR, free ribosomes; LG, lipofuscin granule (residual body); M, mitochondrion; N, nucleus; OS, photoreceptor outer segment; P_1, globular pigment granule; P_2, elongated pigment granule; PS, phagosome; R, rootlet fiber; RER, rough endoplasmic reticulum; SER, smooth endoplasmic reticulum; T, zonula occludens (tight junction); Z, zonula adherens.

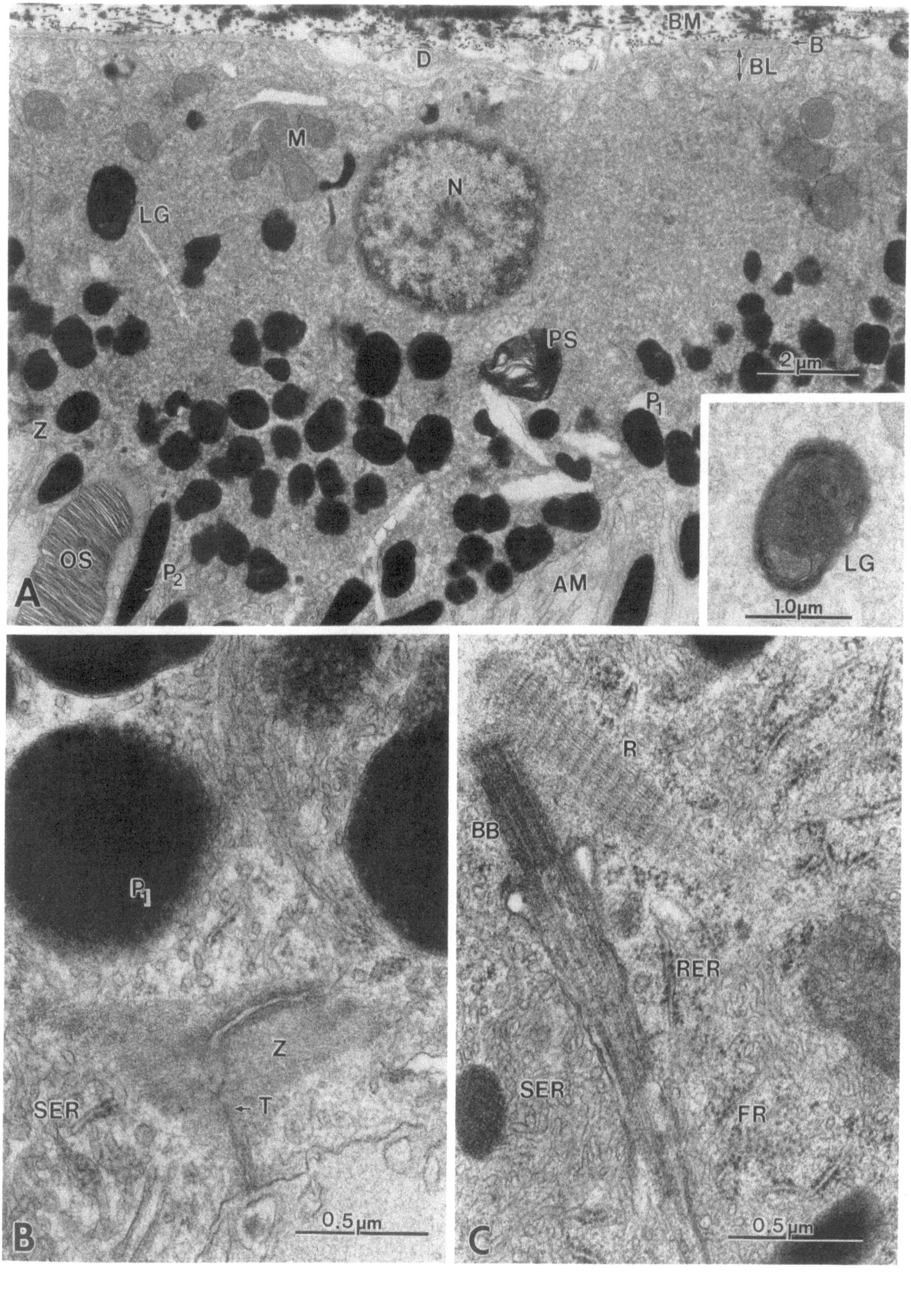
BM
B
D
BL
M
N
LG
PS
2 µm
P1
Z
OS
P2
AM
A
LG
1.0 µm
P1
Z
T
SER
0.5 µm
B
R
BB
RER
SER
FR
0.5 µm
C

Figure 9 shows tangential sections of the basal portion of pigment epithelial cells in a macaque. From Figure 9A it is evident that the basal labyrinth is quite elaborate, although it hardly is recognized in radial sections (see Fig. 8A). A few mitochondria (M) are seen in the microplicae of the labyrinth. Some dark primary lysosomes (L) are seen, and occasionally nuclei are cut at this level. At higher magnification (inset), free ribosomes are recognized in the cytoplasmic extensions.

At a level only slightly deeper in the cells, mitochondria are the predominant feature (Fig. 9B). Primary lysosomes (L) are also seen. Phagosomes at various stages of degradation are scattered in the cytoplasm. Some of them can be designated as lipofuscin granules (LG). The phagosomes (PS) contain membranous material that is derived from rod and cone outer segments. Such bodies commonly are termed multilamellar or myelinoid bodies (Ghadially 1982). The latter term may give rise to confusion, because retinal pigment epithelial cells of several species contain a conspicuous formation of the smooth endoplasmic reticulum named the myeloid body (Kuwabara 1975). This structure was not seen in any of the specimens used to prepare this atlas.

Figure 9. Tangential sections of basal portion of cynomolgus retinal pigment epithelium at parafoveal locus, ×5,460; inset, ×51,250. (A) Basal labyrinth. (B) Mitochondria-rich region. L, primary lysosomes; LG, lipofuscin granule (residual body); M, mitochondrion; N, nucleus; PS, phagosome; R, ribosomes.

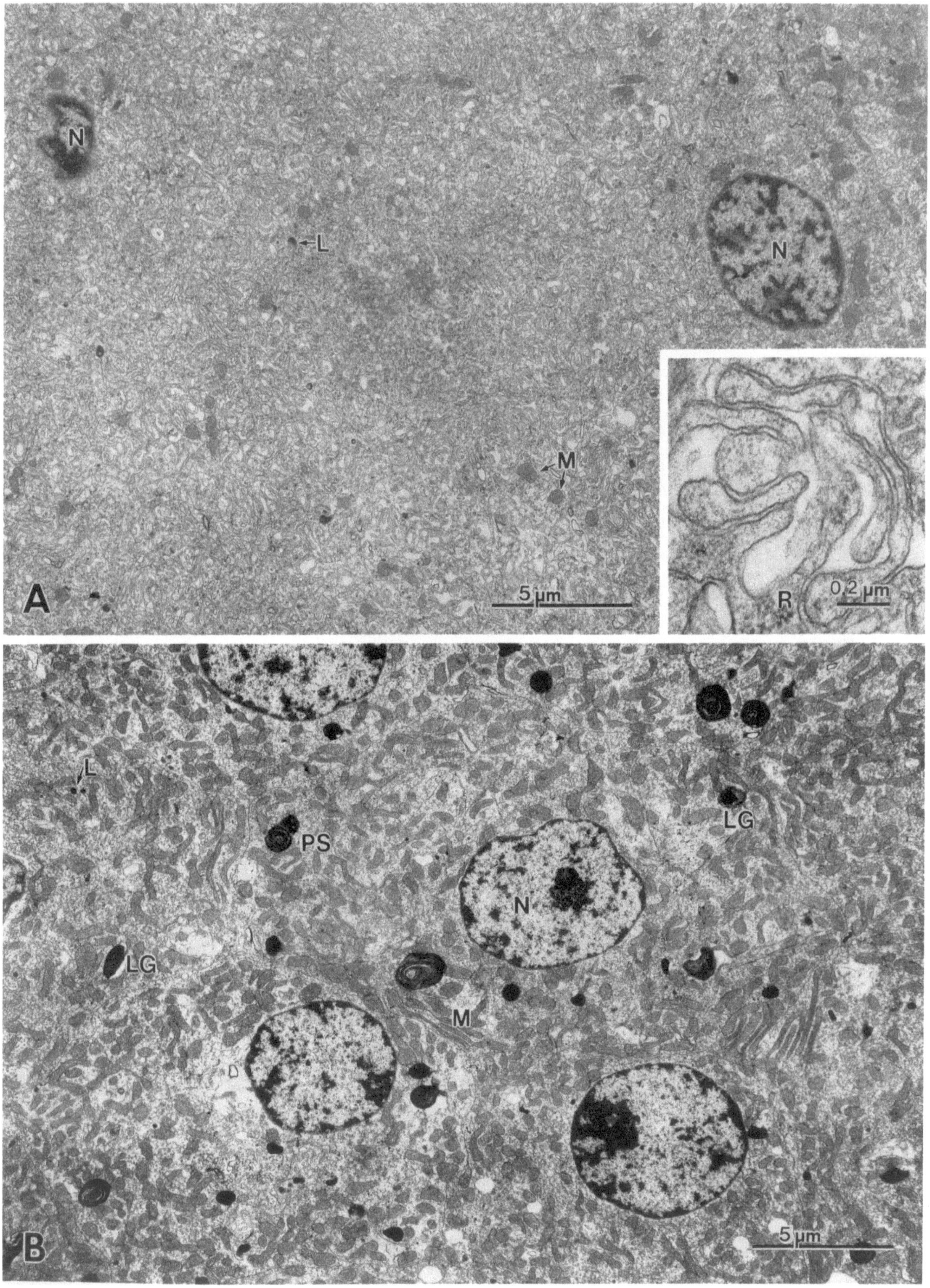
N
L
N
M
5 µm
A
R
0.2 µm
L
PS
LG
N
LG
M
5 µm
B

Figure 10 represents tangential sections through the central region of pigment epithelial cells of the baboon. The profiles of the peripheral cells (Fig. 10A) are significantly larger than those of cells in the foveal region (Fig. 10B). The peripheral cells in Figure 10A are cut slightly deeper than the cells in Figure 9. At this deeper level the mitochondria (M) are aligned along the periphery of the cell. Almost all mitochondria of the retinal pigment epithelial cells are concentrated in the basal portions of the cells near the cytoplasmic membrane.

Some globular pigment granules (P_1) and phagosomes (PS) are seen in Figure 10A. The cytoplasm surrounding the nucleus (N) is filled with tubular smooth endoplasmic reticulum. A few straight flat cisternae of the endoplasmic reticulum (arrowhead) are normally seen at this level of the cells.

At the next level (Fig. 10B) within the pigment epithelial cells, the globular pigment granules predominate in the cytoplasm. Some of them contain dark and light material (arrows). Small stacks of rough endoplasmic reticulum are seen between the pigment granules. This distinct arrangement of the rough endoplasmic reticulum is seen frequently in retinal pigment epithelial cells.

Figure 10. Tangential section through central region of pigment epithelial cells of baboon, × 5,460. (A) Peripheral retina. (B) Fovea; M, mitochondria; N, nucleaus; P_1, globular pigment granule; PS, phagosome; RER, rough endoplasmic reticulum; SER, smooth endoplasmic reticulum; arrows, pigment granules with inhomogeneous content; arrowhead, cisterna of smooth endoplasmic reticulum.

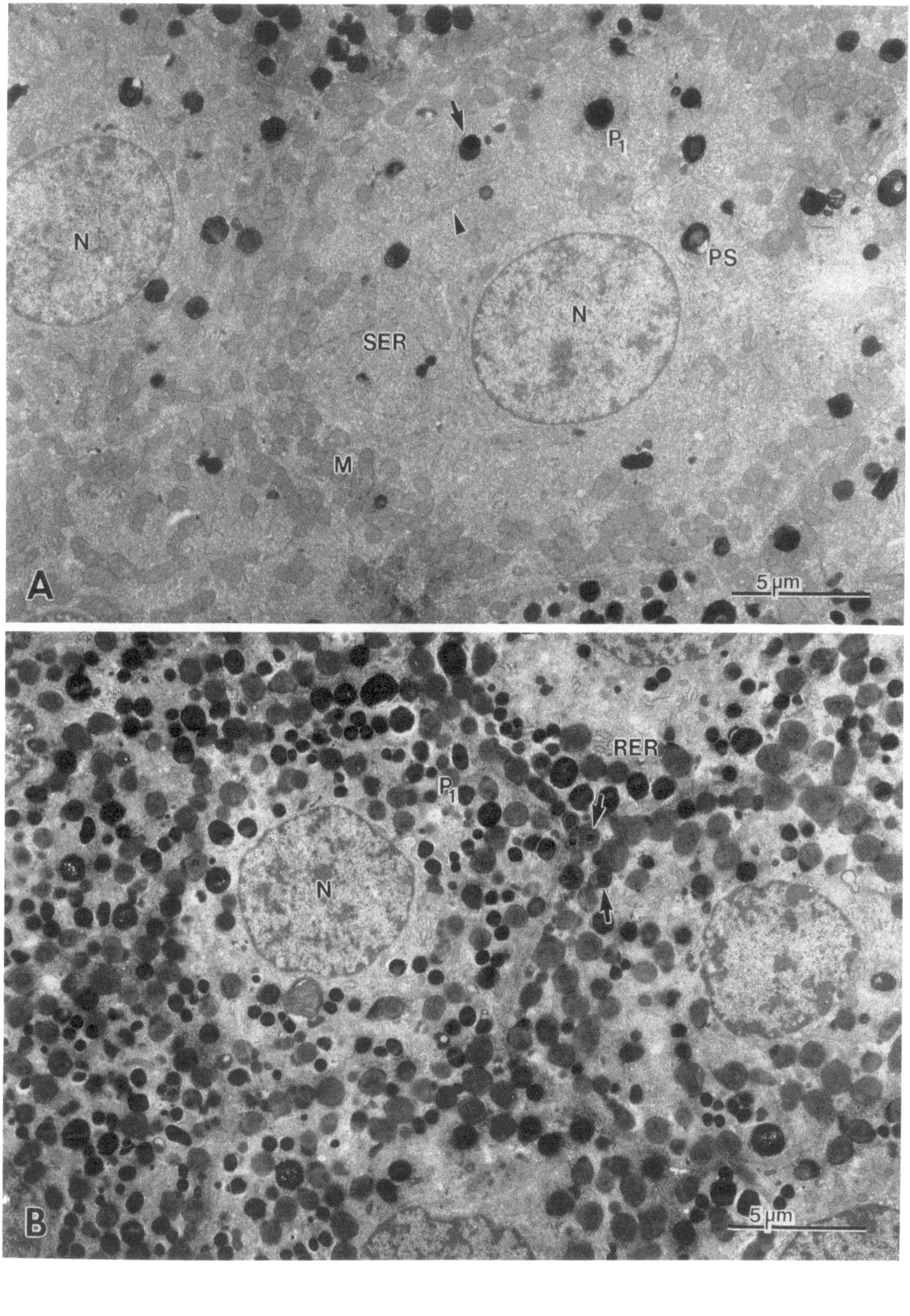
A
N
N
SER
M
P1
PS
5 µm
B
RER
P1
N
5 µm

Figure 11 shows tangential sections through the apical region of the pigment epithelium. The junctional complex of the pigment epithelial cells is depicted in Figure 11A. At this level, the profile of the cells in tangential sections forms very regular hexagons. The plasma membranes of neighboring cells are parallel and are accented by a border of electron-dense material in the cytoplasm. At higher magnification (Fig. 11B) microfilaments can be resolved in this material. The filaments take a course in the plane of the section and parallel to the plasmalemma. There are periodic densities at the inner surface of the plasma membrane (arrows) and within the microfilament bundle (arrowhead).

Microfilaments are equated with actin filaments (Korn 1978). In addition, actin was found to be a major protein in human pigment epithelial cells (Haley et al. 1983), and, in the chicken, actin-containing filaments (Owaribe and Masuda 1982) as well as the actin-associated proteins vinculin and alpha-actinin (Sandig and Kalnins 1988) have been demonstrated in the *zonula adherens* complex of pigment epithelial cells. Thus, if microfilament bundles of the adhering junctions in the pigment epithelium contain actin, the dark periodicities might represent tropomyosin and myosin. In this case the zonules would be capable of actively regulating their tension. The striking resemblance of the zonular filament bundles with "stress fibers," which are involved in cell motility (Geiger 1983), may lend support for this speculation.

In a grazing section of the apex of a pigment epithelial cell (Fig. 11C), the intricate relationship between photoreceptors and pigment epithelium becomes obvious. Tips of rod outer segments are seen in profile, surrounded by microvilli (MV) and microplicae (MP). Many of the microvilli contain elongated pigment granules (P_2) seen in cross section. In the extracellular space an amorphous substance is stained (arrows).

Figure 11. Junctional complex of pigment epithelium. (A) Baboon, fovea, ×5,460. (B) Cynomolgus monkey, fovea, ×13,600. (C) Cynomolgus monkey, parafovea, ×4,620. L, lysosome; MP, microplicae; MV, microvilli; P_2, elongated pigment granule in cross section; RER, rough endoplasmic reticulum; ROS, rod outer segment; SER, smooth endoplasmic reticulum; T, *zonula occludens* (tight junction); Z, *zonula adherens*; solid arrows, subplasmalemmal densities; arrowhead, densities in microfilament bundle; open arrows, amorphous extracellular material.

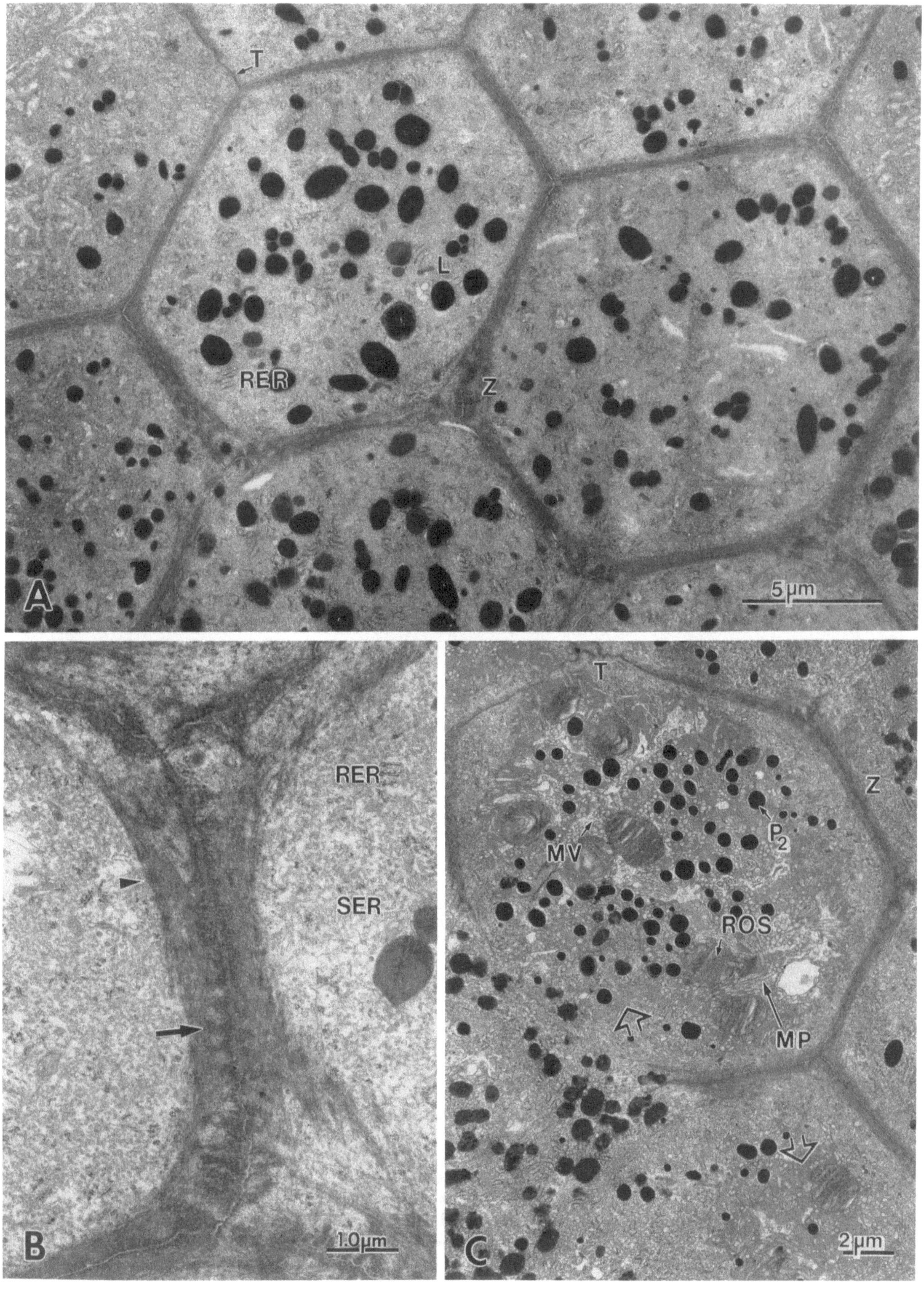
T
L
RER
Z
5µm
A
RER
SER
1.0µm
B
T
Z
P_2
MV
ROS
MP
2µm
C

Figure 12 shows the region of contact between pigment epithelium and rod outer segments in the peripheral retina (Fig. 12A) and in the fovea (Fig. 12B) of the baboon. In the peripheral retina, microvilli (MV) of the retinal pigment epithelium, occasionally containing a pigment granule, are scattered between the tips of rod outer segments. Some rod outer segment profiles are ensheathed by a sleeve of pigment epithelial cytoplasm (inset). These outer segment tips are very likely to be phagocytized (Bok 1985).

Cones are shorter than rods in the peripheral retina. Thus, the pigment epithelial cells must extend long processes to come into contact with cone outer segments. These processes are slender microvilli, many of them containing melanin granules. They are arranged to form a hollow cylindrical structure (C) the diameter of which corresponds to the diameter of cone inner segments. This structure will be called the contact cylinder in the following text. A set of microvilli is in the center of the cylinder. The central microvilli also contain melanin granules. They contact the tip of the cone outer segment which reaches into the cylinder of microvilli. As the foveal center contains only cones, the cylinders formed by pigment epithelial microvilli fuse and become honeycomb like (Fig. 12B).

Figure 12. Tangential section at interface between pigment epithelium and photoreceptor outer segments, baboon. (A) Peripheral retina, ×3,150; inset, ×12,000. (B) Fovea, ×3,150. C, extensions of pigment epithelial cell to contact cone outer segment; COS, cone outer segment; MV, apical microvilli of pigment epithelial cell; P_2, elongated pigment granule in cross section; ROS, rod outer segment; arrows, rod outer segments surrounded by sleeve of pigment epithelial cytoplasm (one is shown in inset); arrowheads, pigment granules between rod outer segments.

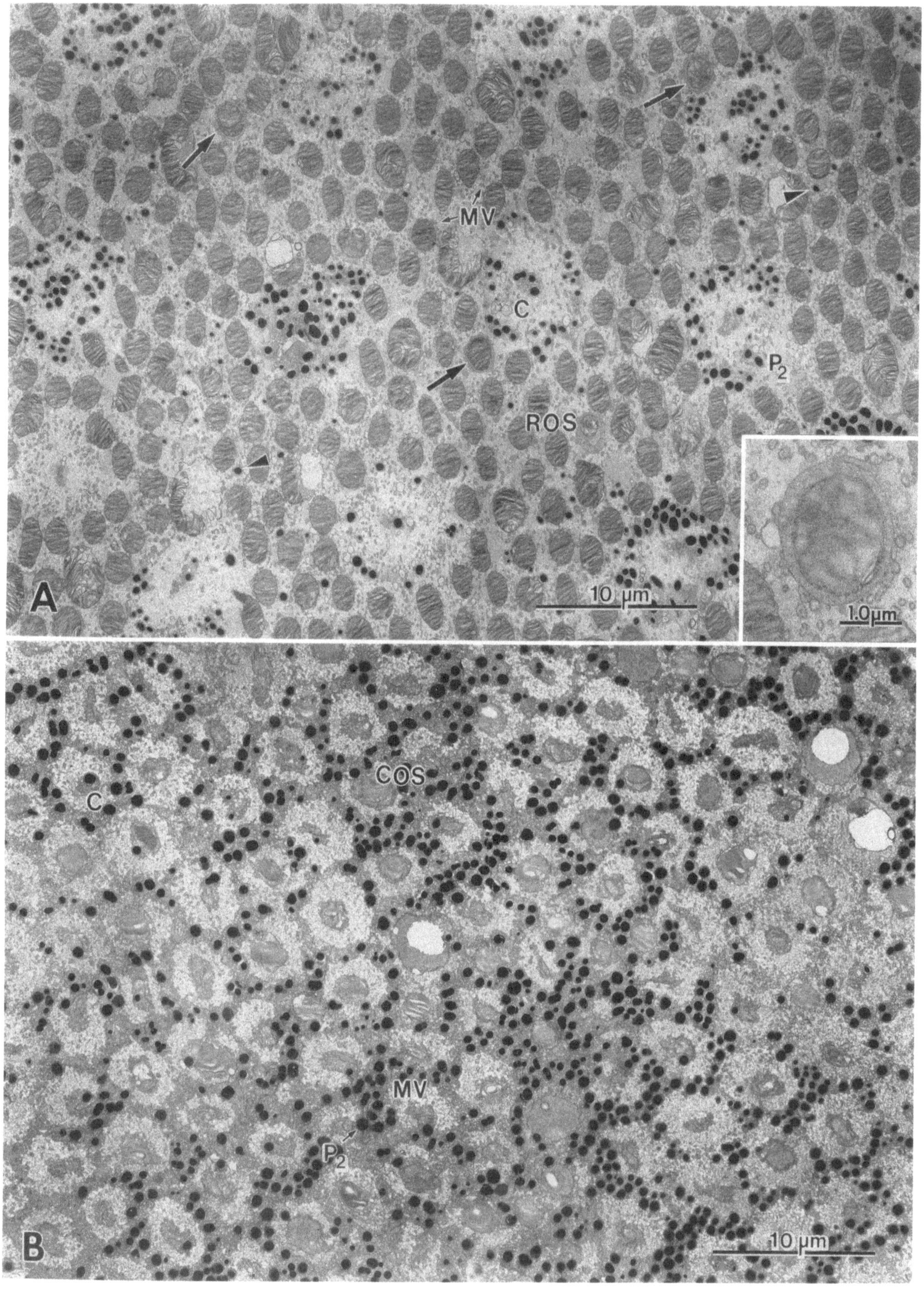
MV
C
P_2
ROS
A
10 µm
1.0µm
COS
C
MV
P_2
B
10 µm

Figure 13 shows cross sections of the contact cylinders at various levels. Proximally, the pigment granules in the microvilli form a complete ring (Fig. 13A). In the center of the contact cylinder is an extension of the pigment epithelial cell with a stack of cone outer segment discs (Fig. 13A). These discs most likely have just been phagocytized. Slightly below this level, the ring of pigment granules in the wall of the contact cylinder is incomplete (Fig. 13B). The central microvilli, some of them containing pigment granules, surround an empty sleeve of pigment epithelial cytoplasm (S).

In Figure 13C only two pigment granules are seen in microvilli of the wall of the contact cylinder. The sleeve (S) in the central accumulation of microvilli is incomplete. The microvilli contain a variety of organelles including mitochondria (M), microtubules, and rough endoplasmic reticulum (Krebs and Krebs 1989a). In the section represented in Figure 13D the wall of the contact cylinder is incomplete. The central sleeve, composed of several subunits, encloses the tip of a cone outer segment. By chance, no disc membranes are in the plane of this section. In the extracellular space, particularly between the microvilli, dark amorphous material is seen.

Photoreceptor Cells

Rod and cone outer segments are the light-absorbing organelles of the photoreceptor cells. Their disc membranes contain the visual pigments. The pigment of rod outer segment discs, rhodopsin, has been well studied biochemically (see Stryer 1988 for details). Rhodopsin molecules take up approximately 50% of the volume of the disc membrane. The disc membrane volume is half of the volume of the rod outer segment (Krebs 1985; Krebs and Kühn 1977). From this, a rhodopsin concentration of 25% results in rod outer segments. Because of this high rhodopsin concentration, rod outer segments are efficient light absorbers. They also are very efficient in generating lighttriggered excitation, a process called visual transduction. Absorbance of a single photon can elicit excitation in mammalian rod cells.

The biochemical events of the transduction process have been discovered (Chader 1985, Stryer 1988, Uhl et al. 1990). By absorption of a photon, rhodopsin becomes activated. It initiates a cascade of enzymatic reactions that lead to a decrease in concentration of cyclic guanosine monophosphate (cGMP) in the interdiscal space of the outer segment (Kühn et al. 1982). This decrease triggers the closing of sodium pores in the plasmamembrane of the outer segment. Rhodopsin remains activated until inactivated by other chemical reactions. The equilibrium between activation of rhodopsin by

Figure 13. Contact cylinders of microvilli contacting tips of cone outer segments, fovea, baboon, ×16,300. (A) Contact cylinder closest to pigment epithelial cell. (B) Slightly further from pigment epithelial cell. (C) Further from pigment epithelial cell. (D) Contact cylinder closest to neuroretina. CM, central microvilli; COS, cone outer segment material; DM, electron-dense amorphous material; M, mitochondrion; P_2, elongated pigment granules in cross section; S, sleeve of pigment epithelial cytoplasm.

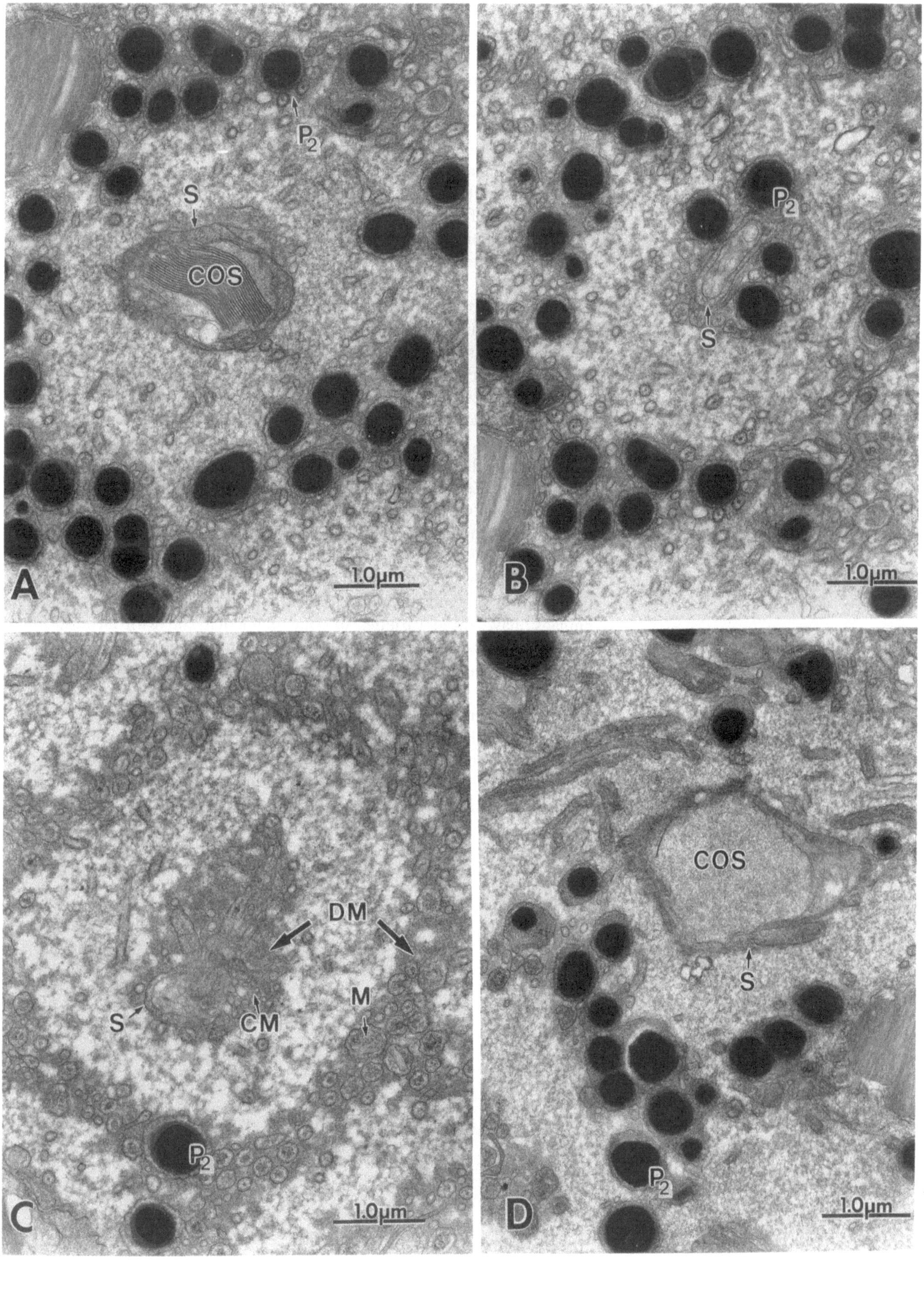

P2
S
COS
1.0µm
A
P2
S
1.0µm
B
DM
S
CM
M
P2
1.0µm
C
COS
S
P2
1.0µm
D

light and its inactivation determines the sensitivity of the system. Calcium plays a role in regulating this equilibrium (Uhl et al. 1990). The hydrolysis of rhodopsin into retinal and opsin has no significance for the transduction process. It occurs within minutes after light absorption and excitation of the rod cell.

In the dark a steady current of sodium ions is generated by ion pumps in the inner segments. This "dark" current enters the outer segments, flows through the cilia into the inner segments, and than back into the subretinal space (Hagins et al. 1970). Depending on the amount of photons absorbed, the intensity of this current is reduced gradually by the light triggered closing of sodium pores in the plasma membrane of outer segments. The reduction of the dark current is the electrophysiological signal of excitation of rod cells.

Whereas rods are able to detect single photons, cones operate at light intensities too intense for rods to function. Thus, there are two separate light-detecting systems in the retina, one for dim and the other for bright light vision. The dim light system is dominated by rods and the bright light system by cones. This duplex system enables the retina to function in the darkness of a starlit night as well as in the 10^{10} fold brightness of the tropical sun at noon.

In primates the cone-dominated bright light system mediates color vision (Gouras 1991; Zrenner 1983). However, the sensation color is not generated in the cone cells. An individual photoreceptor cell is color blind; it merely records the number of absorbed light quanta per unit time without regard for their energy. Color vision is generated by the neurons of the retina.

Primate cones contain visual pigments that differ in their absorption spectra from the rod pigment rhodopsin. Old World primates, including humans, have three varieties of cone pigments: one with an absorption maximum in the red part of the visible light spectrum; another with its maximum in the green; and a third pigment that absorbs maximally in the blue part of the spectrum. The absorption spectra of these pigments overlap considerably, but any cone contains only one of these pigments. The "blue cones," containing the pigment with its maximal absorbance in the blue, are few compared to both of the other cone types. The genes for the human cone pigments have been isolated, and the deduced amino acid sequences are very similar to each other and to the rod pigment (Stryer 1988).

A given light elicits responses of different intensity by each kind of cones. The neural network of the retina generates the sensation of color by comparing these responses. As rods contain a pigment different from that of the cones, they may, in principle, contribute to color vision. With rods fully functioning in dim light, the cone system is shut off. Thus, in this condition there is only one pigment species available and no color vision is possible. At intermediate light levels, however, a contribution of rods to color vision has been demonstrated in humans (Rodieck 1973, pp. 747 ff).

Opsin is a glycoprotein. Its polypeptide chain is synthesized in the rough endoplasmic reticulum of the inner segments. The sugar chain is added in the Golgi apparatus. The molecule remains within the membrane of vesicles and is transported to the apical plasma membrane of the inner segment. The vesicles fuse with the plasmalemma, and rhodopsin is transported along the connecting cilium into the outer segments to be incorporated into the basal disc membranes (Bok 1985). Because of the low viscosity of the lipids in the discs, rhodopsin diffuses freely within these membranes (Liebman and Entine 1974). In rod outer segments, discs are separated from each other and from the plasma membrane. Thus, rhodopsin remains restricted to the disc into which it was inserted. In cone outer segments, however, disc membranes are continuous with the plasma membrane of the outer segments so that the pigment molecules are free to migrate through the whole outer segment.

As mentioned in the previous section, rod and cone outer segments are metabolized by the pigment epithelium. This arrangement makes the photoreceptor cells dependent on the pigment epithelium. The advantage is that discarded disc membranes, which still contain visual pigments, are digested at a location where they do not interfere with the incident light before it can be absorbed by rhodopsin in the outer segments.

Shedding of discs and phagocytosis of the shed membranes by the pigment epithelial cells is a rhythmic activity governed by the light:dark cycle. The large body of literature about this aspect of retinal physiology has been reviewed by Bok (1985). Due to the high concentration of lipids in the stack of disc membranes, the refractive index inside the outer segment differs from that in the extracellular space. The organelle may act as a light guide. The functional significance of this property of photoreceptor outer segments has been analysed by Enoch (1961).

Figure 14 is a drawing of a cone and a rod cell composed of segments of several cells in an electron micrograph of the peripheral baboon retina. The external dimensions and the relative sizes of the cellular domains are drawn to scale, but the cytoplasmic organelles are exaggerated in size.

Cone cells are plumper and shorter than rod cells. The stack of discs fills the outer segment of rods completely; but in cone outer segments, large areas laterally of the disc stacks and often at the tip of the outer segment are free of discs. Cone inner segments contain more mitochondria than rod inner segments do. In both rods and cones, mitochondria are concentrated at the apical ends of the inner segments. This section of inner segments is referred to as ellipsoid. The basal segment, also called myoid, contains the major cytoplasmic organelles, such as smooth and rough endoplasmic reticulum, ribosomes, and, close to the external limiting membrane, Golgi fields.

Because the outer segments are modified cilia, the basal parts of the cilia are well developed. Basal bodies and rootlet fibers are always present.

The perikarya of peripheral cone cells are at the level of the external limiting membrane. The inner cone fiber extends from the perikaryon and ends with the cone cell pedicle at the border to the external plexiform layer. Rod cell perikarya are distributed randomly in the external nuclear layer. Thus, the rod fiber is divided frequently into an outer and an inner segment by the perikaryon. In the fovea, however, there are cone cell perikarya that are not in contact with the external limiting membrane. In these instances cone cells have outer and inner (i.e., Henle's) fibers too.

Figure 14. Drawing of cone cell (left) and rod cell (right) in peripheral baboon retina, composed of sections of several cells in electron micrograph. Outer dimensions are to scale, cytoplasmic organelles are not to scale, ×2,500. C, cilium; CP, cone cell pedicle; DM, membraneous discs; ELM, external limiting membrane; G, Golgi apparatus; IF, inner fiber; IS, inner segment; M, mitochondria; MV, microvilli of pigment epithelial cell; N, nucleus; OF, outer fiber; OS, outer segment; P, perikaryon; PC, *processus calycoides*; PE, pigment epithelium; R, free ribosomes; RER, rough endoplasmic reticulum; RF, rootlet fiber; S, rod cell spherule; SER, smooth endoplasmic reticulum.

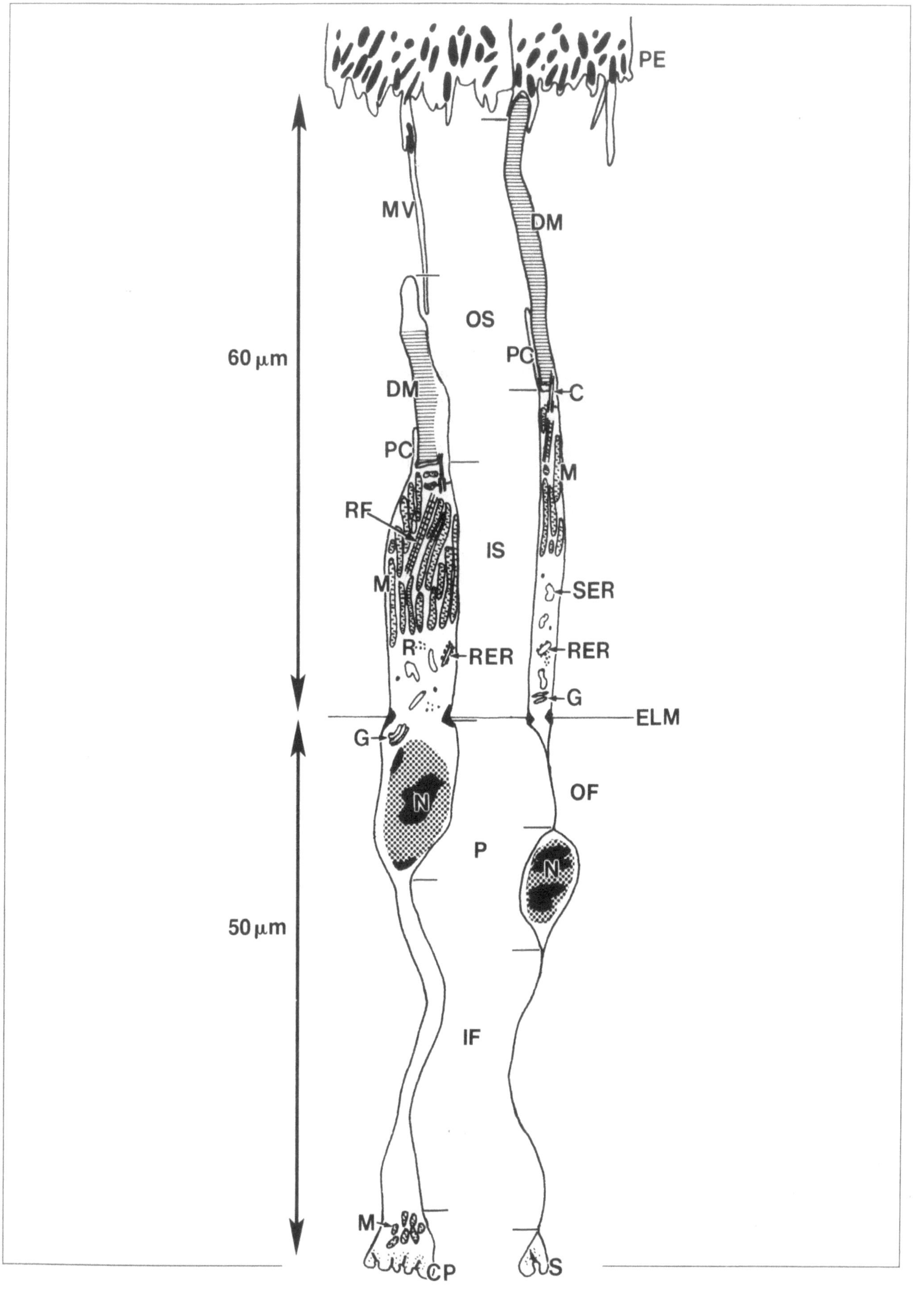
PE
MV
DM
OS
60 μm
PC
DM
PC
C
M
RF
IS
M
SER
R
RER
RER
G
ELM
G
OF
N
P
N
50 μm
IF
M
CP
S

Figure 15 depicts the principal features of a rod. The connecting cilium (CC) is inserted eccentrically in the apex of the inner segment. It originates from a distinct basal body (BB), from which rootlet fibers (RF) extend into the cytoplasm of the inner segment (IS). The basal body has nine triplets of microtubules (Fig. 15B), the typical arrangement for centrioles and basal bodies in mammalian cells.

In a transverse section of a connecting cilium between inner and outer segment (Fig. 15C), the nine double tubules are arranged around the clear center of the connecting cilium. The lack of the central tubules is characteristic of nonmotile sensory cilia. The apical tip of the inner segment (IS) and a few calycoid processes (PC) are also seen in this section (Fig. 15C). In the region of the basal discs, marked by an arrow in Figure 15A, actin has been demonstrated in connecting cilia of the pinfish and rat (Chaitin and Burnside 1989). This is the site of insertion of new discs, and the participation of actin in this process is therefore plausible.

Calycoid processes are straight microvilli originating from the inner segments of rods and cones (Borwein 1981). They extend alongside the basal part of the outer segments, close to the plasma membrane. Their function is not known. Rods have approximately 10 calycoid processes. The outer segment shown in Figure 15D has 9, irregularly distributed along the perimeter of the profile of the outer segment.

The microtubules (MT) of the connecting cilium extend into the outer segment. They course in a column of cytoplasm (CP) with triangular cross section adjacent to the plasmalemma. This column extends almost to the tip of the outer segment and indents the stack of discs. Occasionally a few small flattened membranous vesicles (ME) lie parallel to the plasma membrane. These vesicles resemble displaced fragments of discs.

Figure 15. Rod of baboon. (A) Radial section, ×29,000. (B) Tangential section through apical part of inner segment, ×51,250. (C) Transverse section through connecting cilium, ×71,250. (D) Transverse section through outer segment, ×51,250. BB, basal body; CC, connecting cilium; CP, column of cytoplasm; D, disc membrane; IS, inner segment; M, mitochondrion; ME, disc-like membranes; MT, microtubuli; OM, outer membrane of rod outer segment; OS, outer segment; PC, *processus calycoides;* RF, rootlet fiber; arrows, actin-rich region of connecting cilium.

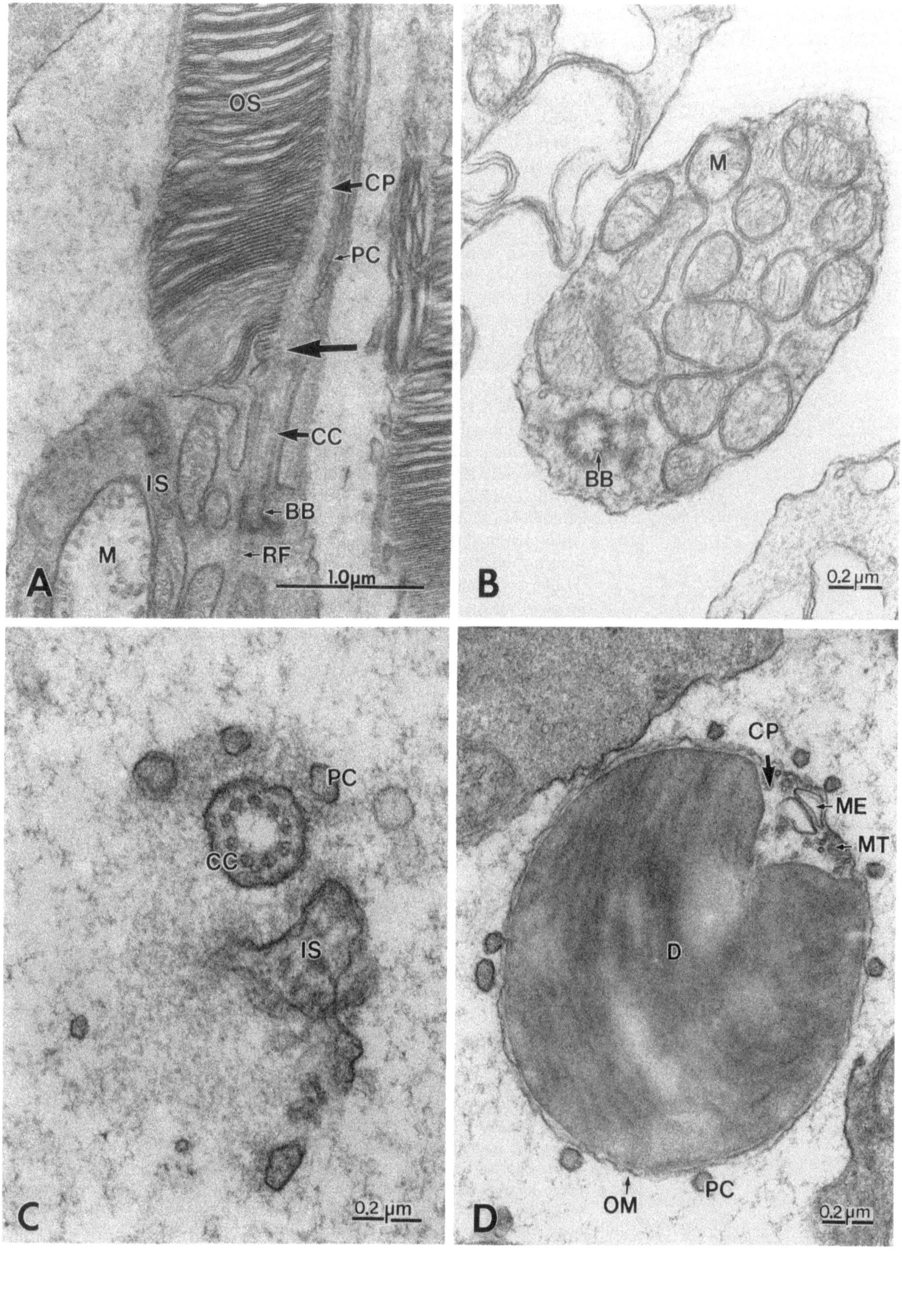
OS
CP
PC
CC
IS
BB
M
RF
1.0μm
A
M
BB
0.2 μm
B
PC
CC
IS
0.2 μm
C
CP
ME
MT
D
OM
PC
0.2μm
D

Figure 16 shows the foveola of the baboon. Rods are absent in the center of the fovea. The foveolar cones are slim and tightly packed; they are 80 μm long, outer and inner segments being each 40 μm long. The diameter of inner segments (IS) is approximately 3 μm and that of the outer segments (OS) is 2 μm. Hogan et al. (1972, p. 493) gave the same dimensions for human foveolar cones.

In the baboon, cytoplasmic inclusions are conspicuous in the cytoplasm of foveolar cone cells. These inclusions have been described as spindle shaped bodies (SB) (Krebs and Krebs 1989b). Their significance is not known, but it has been speculated that they might be connected with the yellow macular pigment. Similar bodies were detected in one cynomolgus retina, but they have been scarce. At higher magnification (see inset of Fig. 16), fibrillar material and electron-dense globules can be resolved within the spindle shaped bodies. The spindle shaped bodies are not enclosed by a membrane.

Cones in the very center of the foveola are packed especially tightly (Ahnelt et al. 1987; Borwein 1981; Krebs and Krebs 1989a). This group of cones, termed the "central bouquet of cones" (Rochon-Duvigneaud 1907) appears in the right part of Figure 16, bordered by the perikaryon in which the nucleus is labeled (N). The nuclei of most of the central cone cells are removed from the external limiting membrane. The part of the cone cells between the perikaryon and external limiting membrane is the outer fiber of the cone cell (OF). It contains cytoplasmic organelles such as free ribosomes, smooth endoplasmic reticulum, and Golgi apparatus (G). The latter is located close to the external limiting membrane.

The external limiting membrane (ELM) is a system of continuous adhering junctions between the apical poles of the radial glial cells (RG) (Müller cells) and between these and the bases of the cones. A 4 μm wide fringe of microvilli (MV), originating from the radial glial cells, penetrates into the subretinal space.

Figure 16. Radial section through foveal center of baboon, ×3,600; inset, ×32,500. ELM, external limiting membrane; G, Golgi apparatus; IS, inner segment; MV, microvilli of radial glial cell; N, nucleus of cone cell; OF, outer fiber of cone cell; OS, outer segment; R, free ribosomes; RG, radial glial cell; SB, spindle shaped body; SRS, subretinal space.

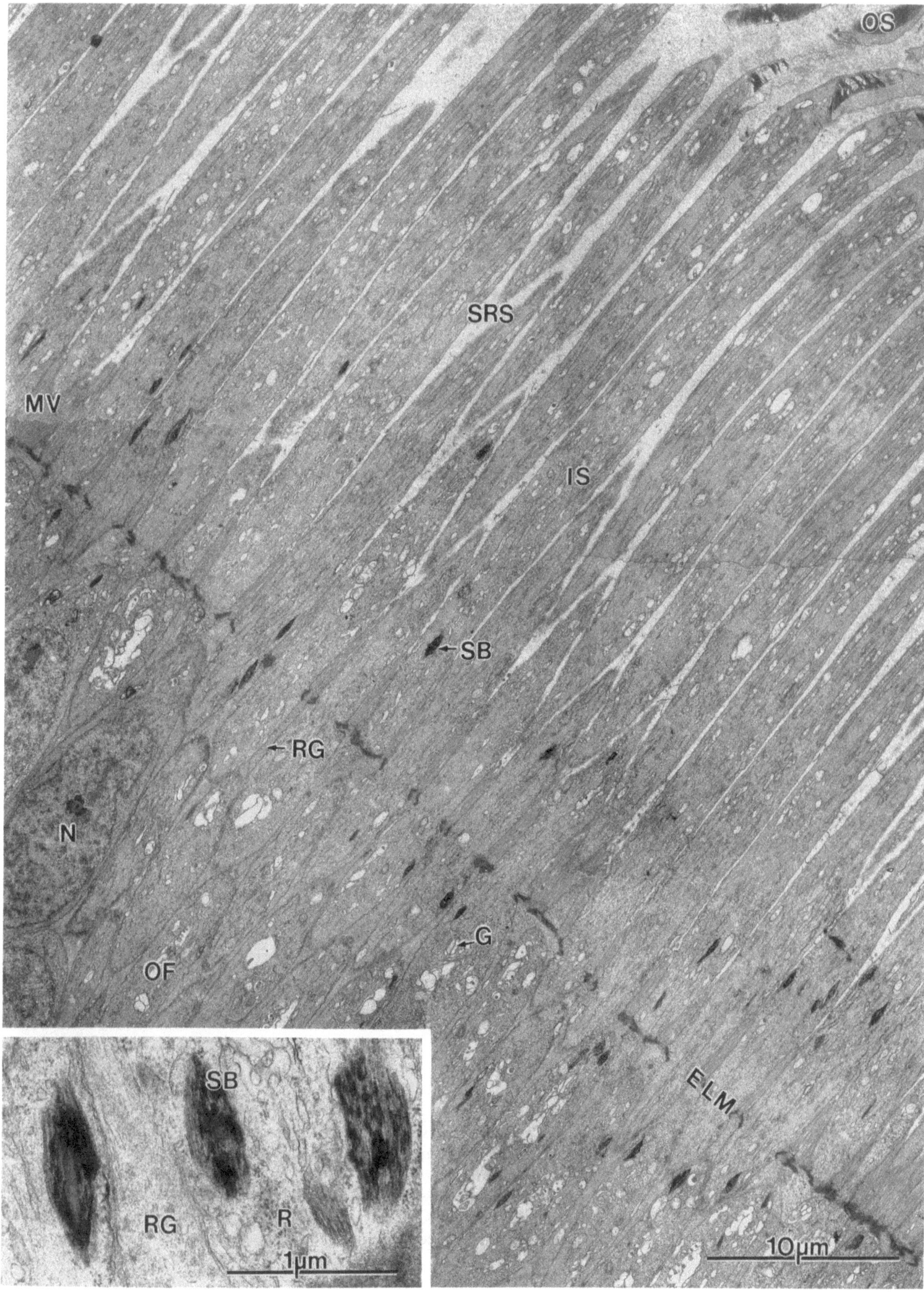
OS
SRS
MV
IS
SB
RG
N
G
OF
ELM
10μm
SB
RG
R
1μm

Figure 17 represents cross sections of foveal cones. In the apical part of the outer segments, the discs (D) frequently are lobulated (Fig. 17A). As in rod outer segments, displaced fragments of disc membranes (ME) occur occasionally in the cytoplasm (CP) adjacent to the outer membrane. A coat of electron-dense material surrounds the outer segments. This material was termed cone matrix sheath (CM). It was seen in the bovine retina (Krebs 1982, p. 24) and has received considerable attention in recent years (e.g., Blanks et al. 1988; Bok 1985; Sameshima et al. 1987). It contains glucoconjugates such as glycolipids, glycoproteins, and proteoglycans, and it may play a role in the interaction between pigment epithelial cells and cones.

Calycoidal processes (PC) of cones are regularly spaced around the basal part of the outer segments. They contain microfilaments and intermediate filaments (Fig. 17C). Continuations between disc membranes and the plasmalemma of cone outer segments are apparent (Figs. 17B, and 17D, arrows).

As in rods, the connecting cilia (CC) do not contain central tubules (Fig. 17E and 17F). The apical pole of inner cone segments contains free ribosomes and a few mitochondrial profiles (M). The basal bodies (BB) of the connecting cilia (CC) are well developed (Fig. 17F).

Cone inner segments occasionally contain displaced synaptic ribbons (SR), complete with a corona of synaptic vesicles (Fig. 17F). It is most probable that these synaptic ribbons have been expressed accidentally in this location and that they are functionally insignificant.

Figure 17. Tangential sections through foveal cones. (A) Apical part of outer segments, cynomolgus monkey, ×20,000. (B,C,D) Basal part of outer segment, baboon. (B) ×27,500. (C) ×77,500. (D) ×51,000. (E,F) Apical poles of inner segments and connecting cilia, baboon. (E) ×27,500. (F) ×22,000. BB, basal body; CC, connecting cilium; CM, cone matrix sheath; CP, cytoplasm; D, disc; M, mitochondrion; ME, disc-like membranes; MT, microtubules; OM, outer membrane; PC, calycoid process; SR, synaptic ribbon; SRS, subretinal space; arrows, continuity between disc and plasmalemma.

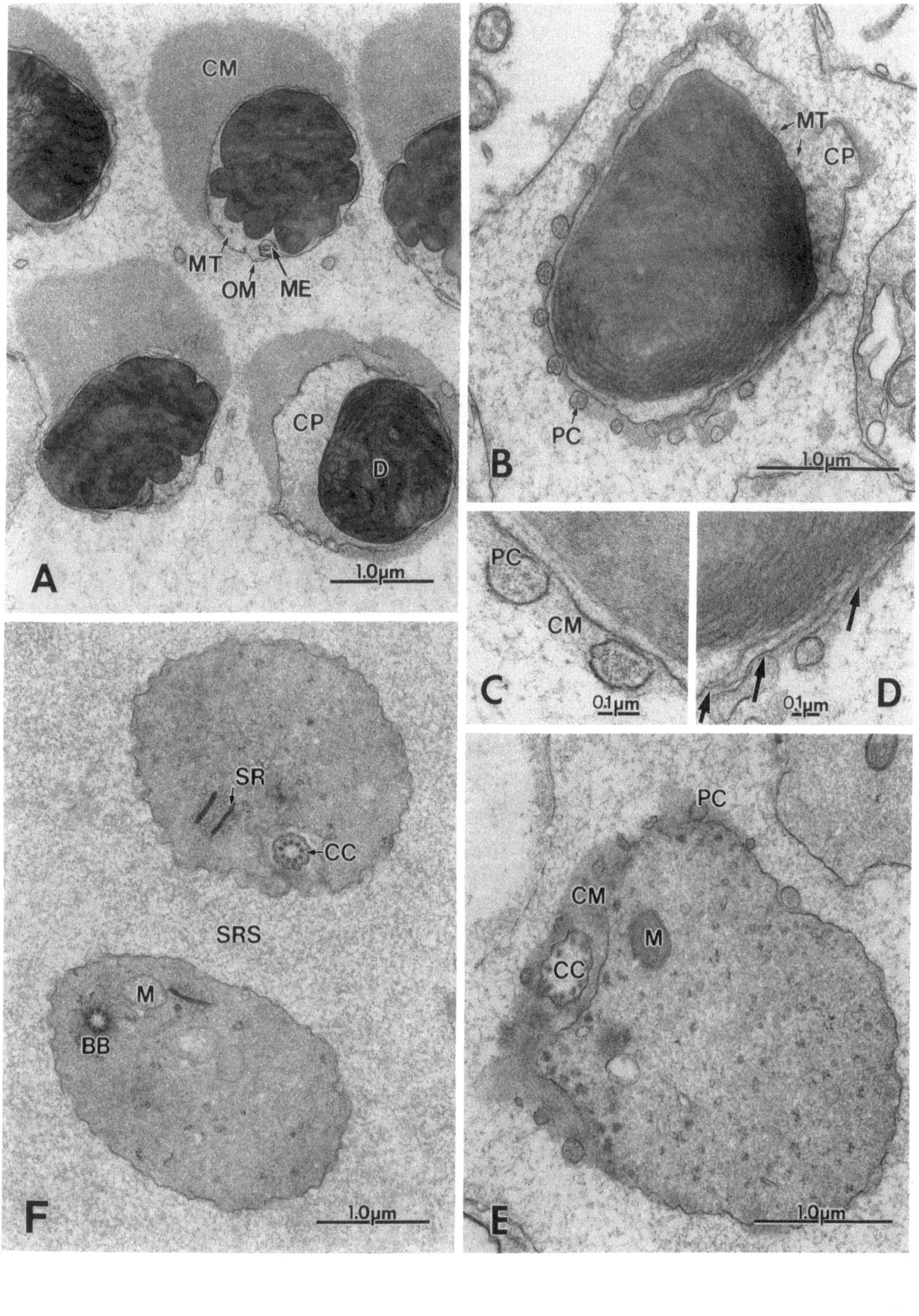
CM
MT
OM
ME
CP
D
A
1.0µm
MT
CP
PC
B
1.0µm
PC
CM
C
0.1µm
0.1µm
D
SR
CC
SRS
M
BB
F
1.0µm
PC
CM
M
CC
E
1.0µm

Figure 18 shows the mosaic of rods and cones of the baboon in three retinal loci. The tangential sections were cut through the mitochondria-rich portion of inner segments. In the peripheral retina (Fig. 18A), the center-to-center spacing between cones (C) is 10 to 15 μm; between rods (R), it is 3 to 5 μm. Cone inner segments appear darker than those of the rods, mainly because of their electron-dense, tightly packed mitochondria.

In the parafoveal sample (Fig. 18B), the cones are closer together, 8 to 10 μm center-to-center, but the rod inner segments have the same spacing as those of the peripheral retina. There is a larger empty extracellular space around cones than around rods.

In the central fovea (Fig. 18C), the cones are tightly packed in a hexagonal pattern. The extracellular space is reduced to a cleft of 20 to 50 nm in width. The center-to-center spacing is approximately 4 μm, about equal to the diameter of the inner segments. A few rods are scattered between the cones. These rods are extremely thin, about 1 μm in diameter. They are absent only from the central part of the fovea, which is formed by the "central bouquet of cones" (Rochon-Duvigneaud 1907). The mitochondria in the foveal cones are less electron-dense than those of the peripheral cones and are not as tightly packed.

Figure 18. Tangential section through peripheral (A), parafoveal (B), and foveal (C) sample of baboon retina at level of rod and cone ellipsoids, ×3,200. C, cone inner segment; R, rod inner segment; SRS, subretinal space.

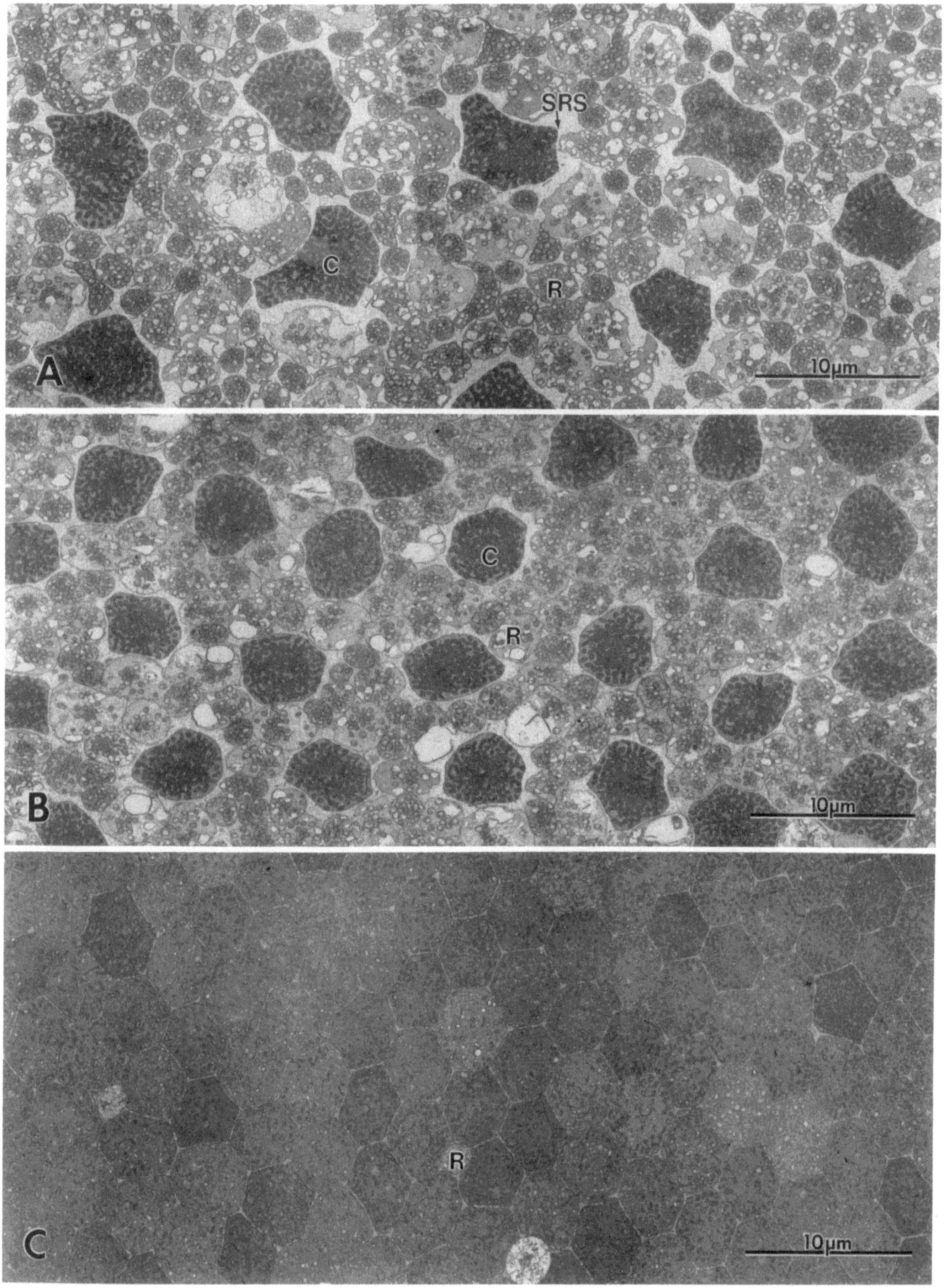
SRS
C
R
A
10μm
C
R
B
10μm
R
C
10μm

Figure 19 depicts some details of the ellipsoid region of rod (R) and cone (C) inner segments. In longitudinal sections of cone and rod inner segments, the slender mitochondria can be seen tightly packed in parallel. In cones they can be as long as 14 μm. In the apical pole of cone inner segments, the tips of the mitochondria are bent into a direction almost perpendicular to the long axis of the cone (Fig. 19A, asterisk). These tips are cut perpendicularly in radial sections, giving the false impression of a separate group of short mitochondria in the apex of cone inner segments (arrow). The mitochondria in cones are more electron-dense than those of rods, partly because of their higher concentration of crista membranes, and partly because of the higher density of their interior matrix.

In cross section, the mitochondria in rods are circular and 0.20 to 0.25 μm in diameter (Fig. 19B). The cone mitochondria are either round or flattened and are 0.20 to 0.25 μm thick. Cone inner segments are surrounded by a 0.25- to 1-μm-wide cell-free section of the subretinal space (SRS).

Striated rootlet fibers (SR) are most prominent in rod inner segments of the pygmy marmoset (Fig. 19C). They have a backbone of microfilaments and show a regular cross striation with a periodicity of 62 nm. The fiber in Figure 19C is tripartite. As seen in this example, mitochondria (MR) are closely associated with the rootlet fiber. This arrangement suggests that the rootlet fiber is an energy-consuming structure.

Figure 19. Ellipsoid region of rod and cone inner segments. (A) Radial section, parafovea, baboon, ×5,700. (B) Tangential section, parafovea, baboon, ×10,600. (C) Tangential section, peripheral retina, pygmy marmoset, ×37,000. C, cone; M, mitochondrion; MR, mitochondrion adjoining striated rootlet fiber; R, rod; SR, striated rootlet fiber; SRS, subretinal space; arrow, short mitochondrial profiles in apical pole of cone inner segment; asterisk, bend in cone mitochondria.

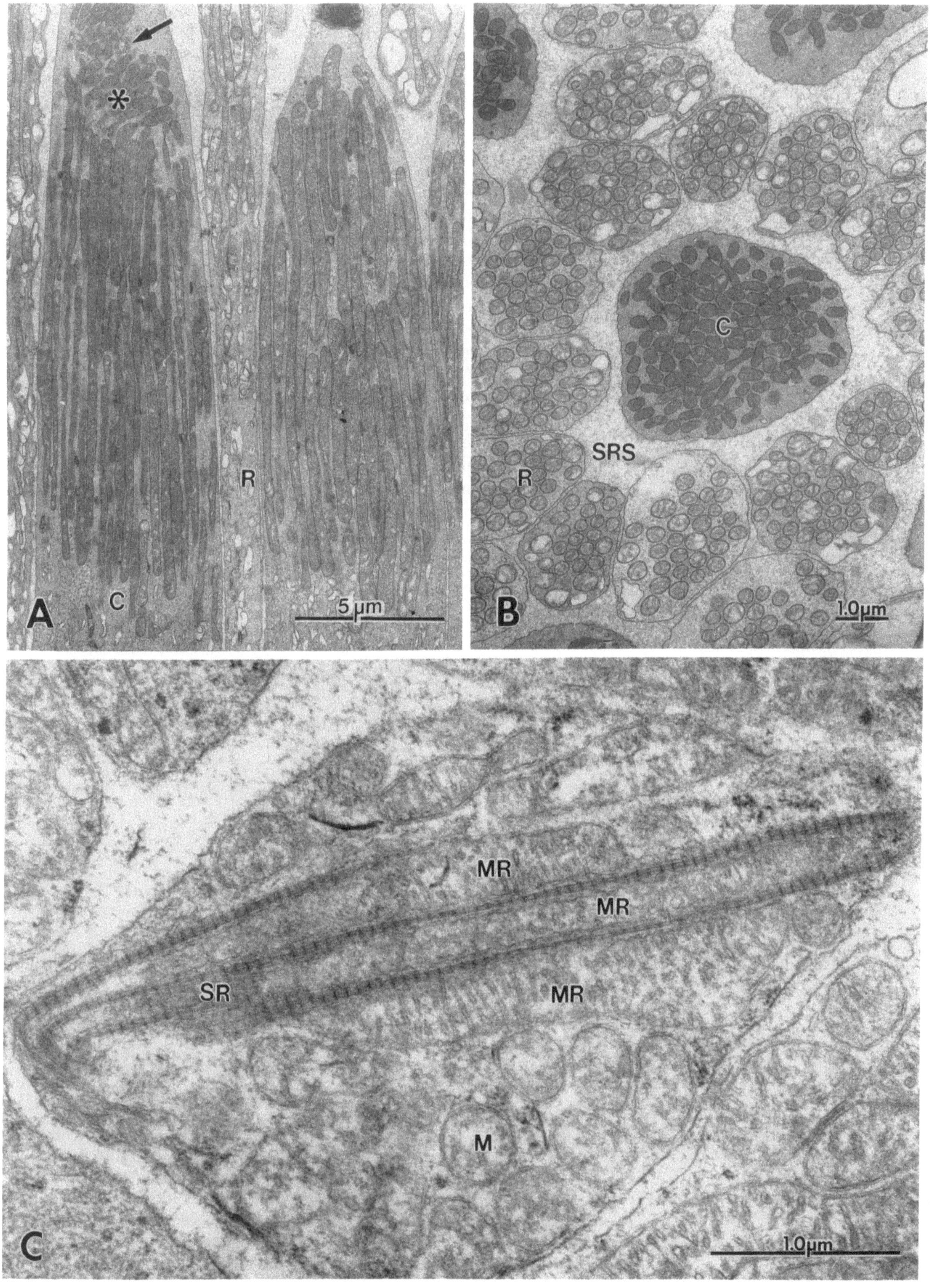

A
C
R
5 μm
B
C
SRS
R
1.0μm
C
MR
MR
SR
MR
M
1.0μm

Figure 20 shows the basal region of rod (R) and cone (C) inner segments. The foveal cone inner segments, about halfway between ellipsoid and external limiting membrane, contain small stacks of rough endoplasmic reticulum (RER), collapsed *cisternae* of smooth endoplasmic reticulum (SER) and abundant microtubules (MT), oriented longitudinally. In the baboon, cross sections of the spindle-shaped bodies (Fig. 16: SB) are seen occasionally.

Close to the external limiting membrane, in the region of the microvilli (MV) of the radial glia, rod (R) and cone (C) inner segments contain most of their cytoplasmic organelles. There are some lysosomes (L), occasional mitochondria (M), free ribosomes (FR), rough (RER) and smooth endoplasmic reticulum (SER), microtubules (MT), and Golgi apparatus (G). A bundle of microfilaments associated with cisternae of the smooth endoplasmic reticulum (SR) is regularly seen; it is the cross section of the striated rootlet fiber. The association of the rootlet fibers with endoplasmic reticulum cisternae has been described as the "rootlet vacuole system" by Cohen (1961).

Cytoplasmic extensions of rod inner segments penetrate into grooves of adjacent rods and cones (Fig. 20B: CZ). Usually there is a cisterna of the endoplasmic reticulum close to this contact zone. No specialization of the plasmalemma is seen in these contact zones. Similar connections have been described in human parafoveal photoreceptors by Ringvold and Davanger (1989).

Figure 20. Tangential section through myoid region of inner segments. (A) Fovea of baboon, ×11,200; inset: details of cone inner segment cytoplasm, ×64,000. (B) Peripheral retina of cynomolgus monkey close to external limiting membrane, ×35,000. C, cone; CZ, contact zone between rod inner segments; FR, free ribosomes; G, Golgi apparatus; L, lysosome; M, mitochondrion; MT, microtubules; MV, microvilli of radial glia; RER, rough endoplasmic reticulum; SB, spindle shaped body; SER, smooth endoplasmic reticulum; SR, striated rootlet fiber.

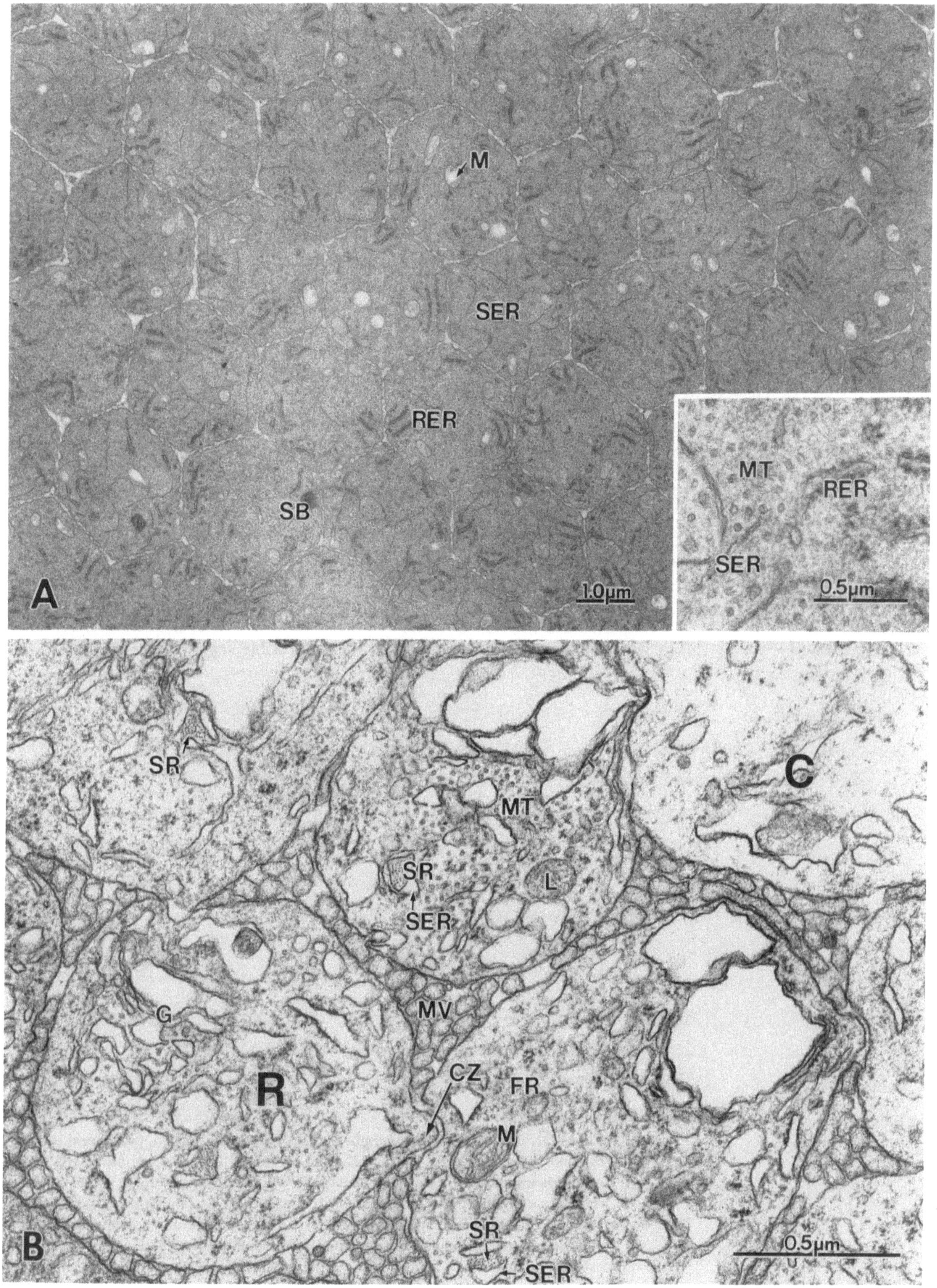
M
SER
RER
SB
A
1.0μm
MT
RER
SER
0.5μm
SR
MT
C
SR
L
SER
G
MV
R
CZ
FR
M
SR
SER
B
0.5μm

As seen in **Figure 21**, the external limiting membrane is a system of continuous adhering junctions (arrowheads) between adjacent cytoplasmic membranes of apices of the radial glial cells (RG) and of rod (R) and cone (C) bases. It is the internal border of the subretinal space. It also delineates the level to which the retinal radial glial cells extend. As mentioned previously, the external limiting membrane is at the level of the cytoplasmic domain of the Golgi apparatus (G) in the photoreceptor cells.

As generally assumed of adhering junctions (Staehelin 1975), the function of the external limiting membrane is primarily mechanical. It glues the photoreceptor cells into a fixed position so that each one of them samples signals from a strictly defined point in the plane of the retina onto which the image of the outside world is projected. It also has been suggested that the external limiting membrane plays a major role as a diffusion barrier between subretinal space and the extracellular compartment of the external nuclear layer. However, the recent discovery of large pores in the external limiting membrane of the central fovea (Fig. 21B) suggested that the limited pore size of the external limiting membrane is not essential for retinal function (Krebs I. and Krebs 1989).

A discontinuity of the external limiting membrane in the central human fovea is shown in Figure 21B. The arrows point to clefts between cone bases connecting subretinal space and the extracellular compartment of the external nuclear layer. The subretinal space is characterized by the microvilli of the radial glia (at SRS in Fig. 21B). In the external nuclear layer, the space between the cone cells is filled by electron-dense cytoplasm of radial glial cells (RG). In Figure 21B, it is possible to trace the extracellular space, without crossing an adhering junction, from the subretinal space side of the adhering zonules, along the arrows and between the radial glial cell cytoplasm and cone cells into the external nuclear layer. This situation cannot be mimicked by any possible angle of sectioning.

Figure 21. External limiting membrane. (A) Tangential section, peripheral retina, cynomolgus monkey, ×3,800. (B) Human fovea, ×7,000. C, cone basis; CP, cone cell perikaryon; ENL, external nuclear layer; G, Golgi apparatus; R, rod basis; RF, rod cell outer fiber; RG, radial glial (Müller) cell; SER, smooth endoplasmic reticulum; SRS, subretinal space; arrow, clefts between cone inner segments without adhering zonule, connecting subretinal space and extracellular compartment of external nuclear layer; arrowheads, *zonula adherens* of external limiting membrane.

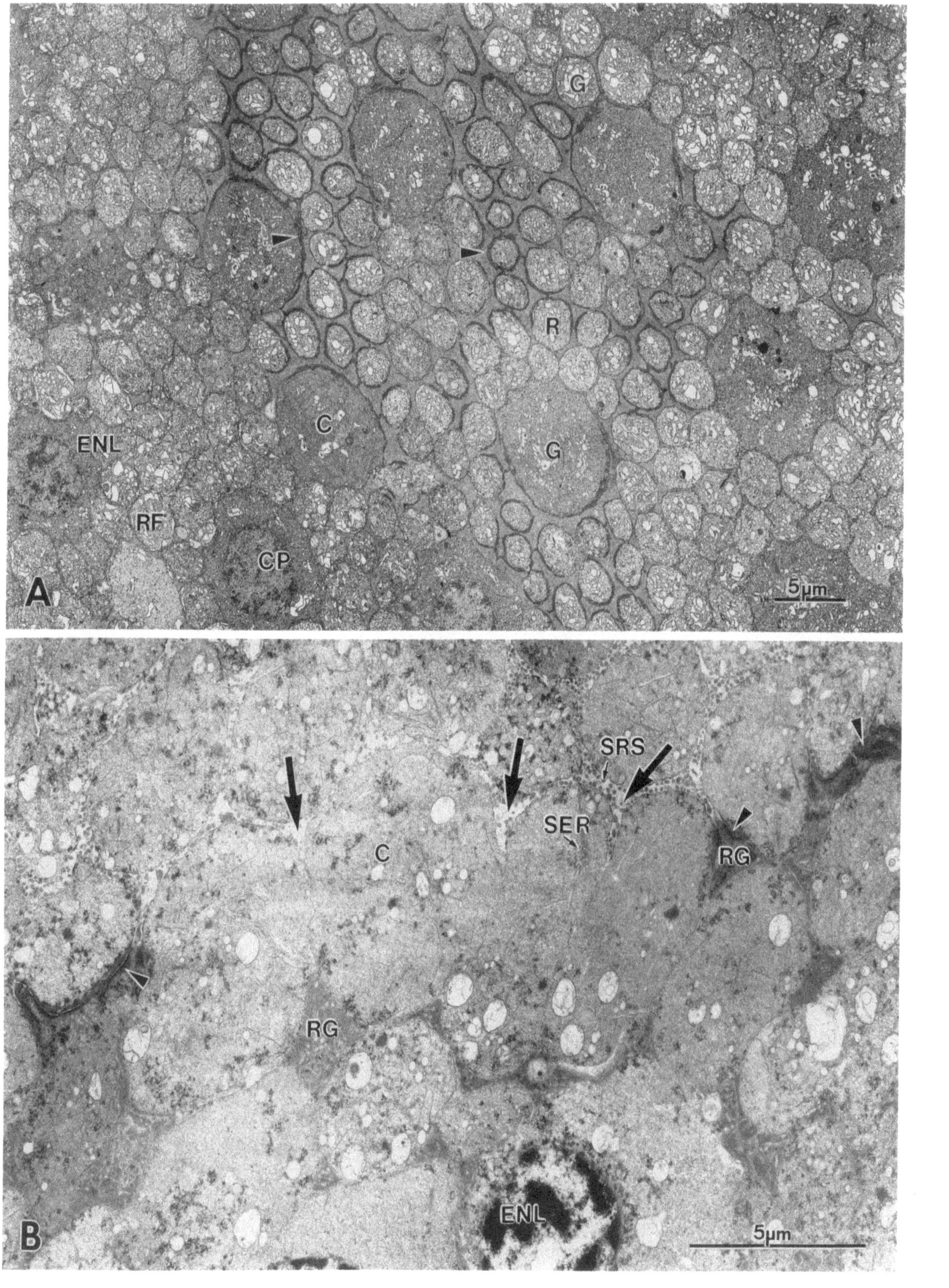
G
R
C
G
ENL
RF
CP
A
5µm
SRS
SER
C
RG
RG
ENL
B
5µm

In **Figure 22**, some details of the layer of the external limiting membrane are shown. The radial glial cells (RG) tightly fill the space between the bases of rods (R) and cones (C) (Fig. 22A and 22B). The apical extension of the radial glial cell in Fig. 22A is in contact with one cone and eight rods with which it is connected by adhering zonules. The Golgi fields (G), which dominate this region of the photoreceptor cells, are apparent. In the cone inner segment, a spindle shaped body (SB) is cut transversely. The cytoplasm of the radial glial cell, at the level depicted in Figure 22, is rich in cytoplasmic organelles, intermediate filaments, and glycogen. Most prominent are the adhering junctions (ZA). They closely resemble the *zonulae adherentes* of the pigment epithelium (Figs. 8 and 11). At in those, microfilament bundles (MF) run parallel to the cytoplasmic membrane; periodic densities (arrows) are present, and electron-dense particles (arrowheads) decorate the inner surface of the plasmalemma. These structural features suggest the presence of actin and actin-associated proteins. Thus, it might be possible that the external limiting membrane is able actively to regulate its tension. The histochemical demonstration of an actin-like immunoreactivity in the external limiting membrane of the developing retina in mice supports this hypothesis (Chaitin et al. 1988).

Solitary cilia occur in the apical extensions of retinal radial glial cells (Fig. 22D). They have a well developed basal body (BB) and a short rootlet fiber (RF). The rootlet fiber of a rod cell is seen in cross section in Figure 22D (RV). It is associated with *cisternae* of the smooth endoplasmic reticulum (see also Fig. 20).

The fringe of microvilli (MV), extending into the subretinal space from the apices of radial glial cells, is seen in Figure 22E. The microvilli are densely packed, straight, and approximately 4 μm long. They are embedded in a matrix of electron-dense material. It has been shown that they have an unusually high density of Na^+, K^+ pump sites in their membranes (Reichenbach and Eberhardt 1988), suggesting that they contribute to the control of the ionic equilibrium within the subretinal space. Between foveal cones, an accumulation of electron-dense extracellular material (asterisk) is seen in the region of the subretinal space bordering the external limiting membrane. The significance and chemical composition of this material are not known.

Figure 22. Layer of external limiting membrane. (A) Tangential section, peripheral retina, cynomolgus monkey, ×9,700. (B) Outline of profile of radial glial cell in Fig. 22A. (C) *Zonula adherens*, peripheral retina, cynomolgus monkey, ×31,100. (D) Basal body and striated rootlet fiber in radial glial cell, peripheral retina, cynomolgus monkey, ×38,800. (E) Radial section, fovea, baboon, ×13,700. BB, basal body; C, cone inner segment; FR, free ribosomes; G, Golgi apparatus; MF, microfilaments; MV, microvilli; R, rod inner segment; RF, rootlet fiber; RG, radial glial cell; RV, rootlet vacuole system; SB, spindle shaped body; SRS, subretinal space; ZA, *zonula adherens*; arrows, dense material in microfilament bundle of adhering zonule; arrowhead, dense material at cytoplasmic membrane in adhering zonule; asterisk, electron-dense material in subretinal space.

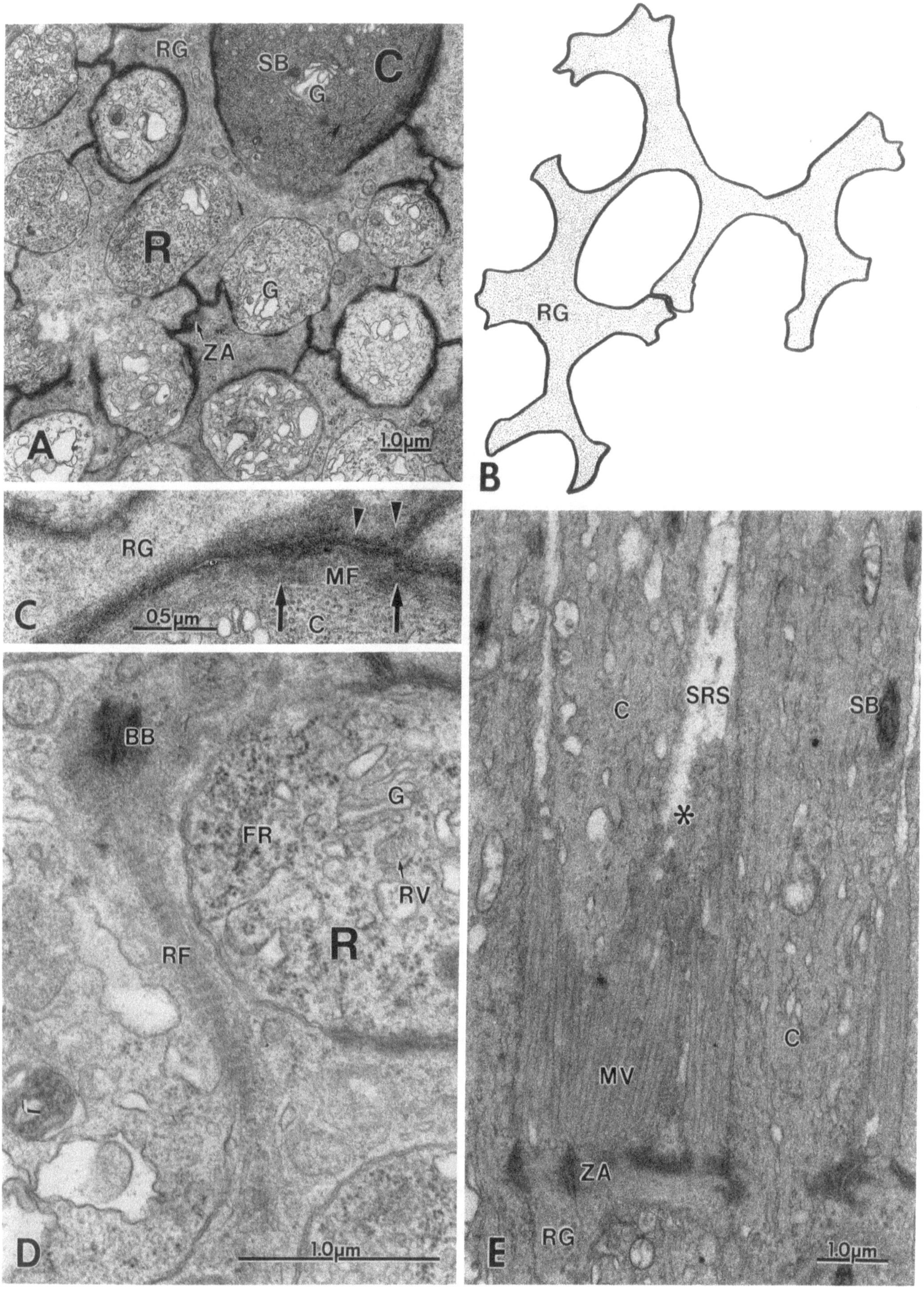
RG
SB
C
G
R
G
ZA
A
1.0µm
RG
B
RG
MF
C
0.5µm
C
BB
G
FR
RV
R
RF
D
1.0µm
C
SRS
SB
*
C
MV
ZA
RG
E
1.0µm

Figure 23 shows tangential sections through the external nuclear layer. It is dominated by profiles of the photoreceptor cell perikarya (Fig. 23A). Except in the foveal area, only rod cell nuclei (R) are cut in a tangential section through the center of this retinal layer. Between the rod cell perikarya are cross sections of rod and cone cell fibers. The inner fibers contain microtubules as their dominating cytoplasmic constituent (Fig. 23B). Thus, they may be considered axons. Inner rod cell fibers (RF) are approximately 2 μm in diameter and contain approximately 20 microtubules; cone cell fiber profiles (CF) are 14 to 20 μm wide and contain 500 to 600 microtubules. The rod and cone cell fibers are completely insulated from each other and from the perikarya by the radial glial cell cytoplasm, which contains only a few cytoplasmic organelles.

Figure 23. Tangential section through external nuclear layer. (A) Peripheral retina, cynomolgus monkey, ×5,000. (B) Parafoveal retina, baboon, ×51,250. CF, inner cone cell fiber; FR, free ribosomes; M, mitochondrion; R, rod cell; RF, inner rod cell fiber.

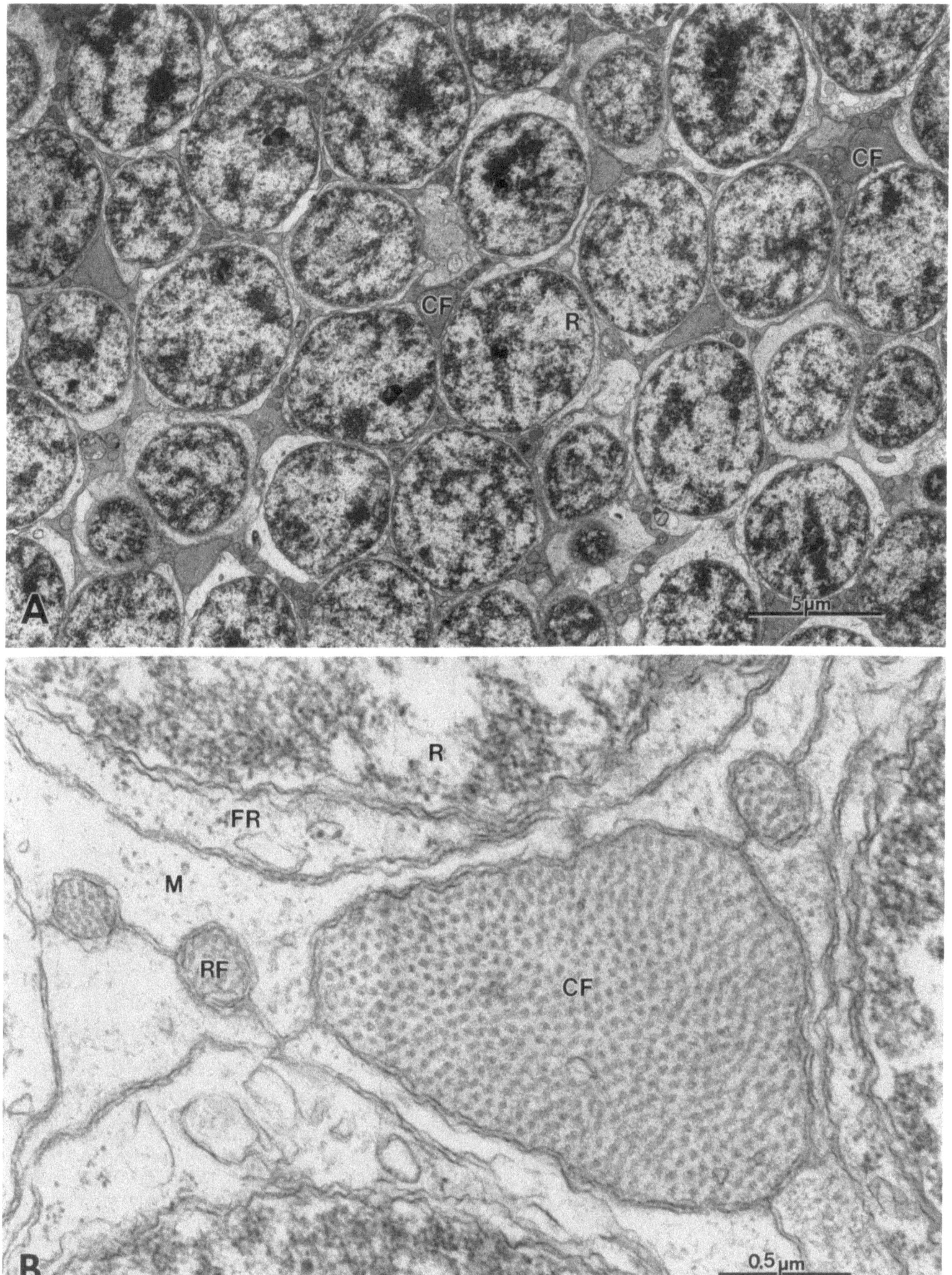
CF
CF
R
A
5 µm
R
FR
M
RF
CF
B
0.5 µm

Figure 24 depicts radial sections of cone cell pedicles (C) and rod cell spherules (R) of the baboon. The cone cell pedicle in Figure 24A is 7 to 8 μm wide at its base. It contains a cluster of mitochondria at the transition of inner cone cell fiber to pedicle. The mitochondria are 0.6 μm in diameter. Synaptic contacts to horizontal and bipolar cells are made by several invaginated ribbon synapses (IS) at the plate-like base of the pedicle. A single cone has 15 to 20 synaptic ribbons (SR).

Rod cell spherules are approximately 3 μm in diameter. They have usually one synaptic ribbon synapse. Mitochondria are few in rod cell spherules.

Figure 24B is an example of an invaginated ribbon synapse in a cone cell pedicle. The cytoplasmic space of the pedicle is filled with clear synaptic vesicles (SV). A conglomerate of tubules (T), as seen adjacent to the synaptic ribbon, is frequently encountered in cone cell pedicles. It appears as if these tubules originated from fused vesicles.

Synaptic ribbons (SR) are characteristic constituents of synaptic terminals

Figure 24. Radial section through rod and cone cell terminals, baboon. (A) Cone cell pedicle and rod cell spherules, ×8,200. (B) Invaginated synapse of cone cell pedicle, ×51,250. (C) Conventional (flat) synapse in cone pedicle, ×75,000. (D) Contact between rod cell spherule and cone cell pedicle, ×51,250. B, bipolar cell process; C, cone cell pedicle; H, horizontal cell process; IS, invaginated synapse; J, junction between rod cell spherule and cone cell pedicle; M, mitochondrion; R, rod cell spherule; RG, radial glial cell; S, conventional (flat) synapse; SR, synaptic ribbon; SV, synaptic vesicles; T, tubular structures; arrow, arciform density; arrowhead, dense material between horizontal cell terminals.

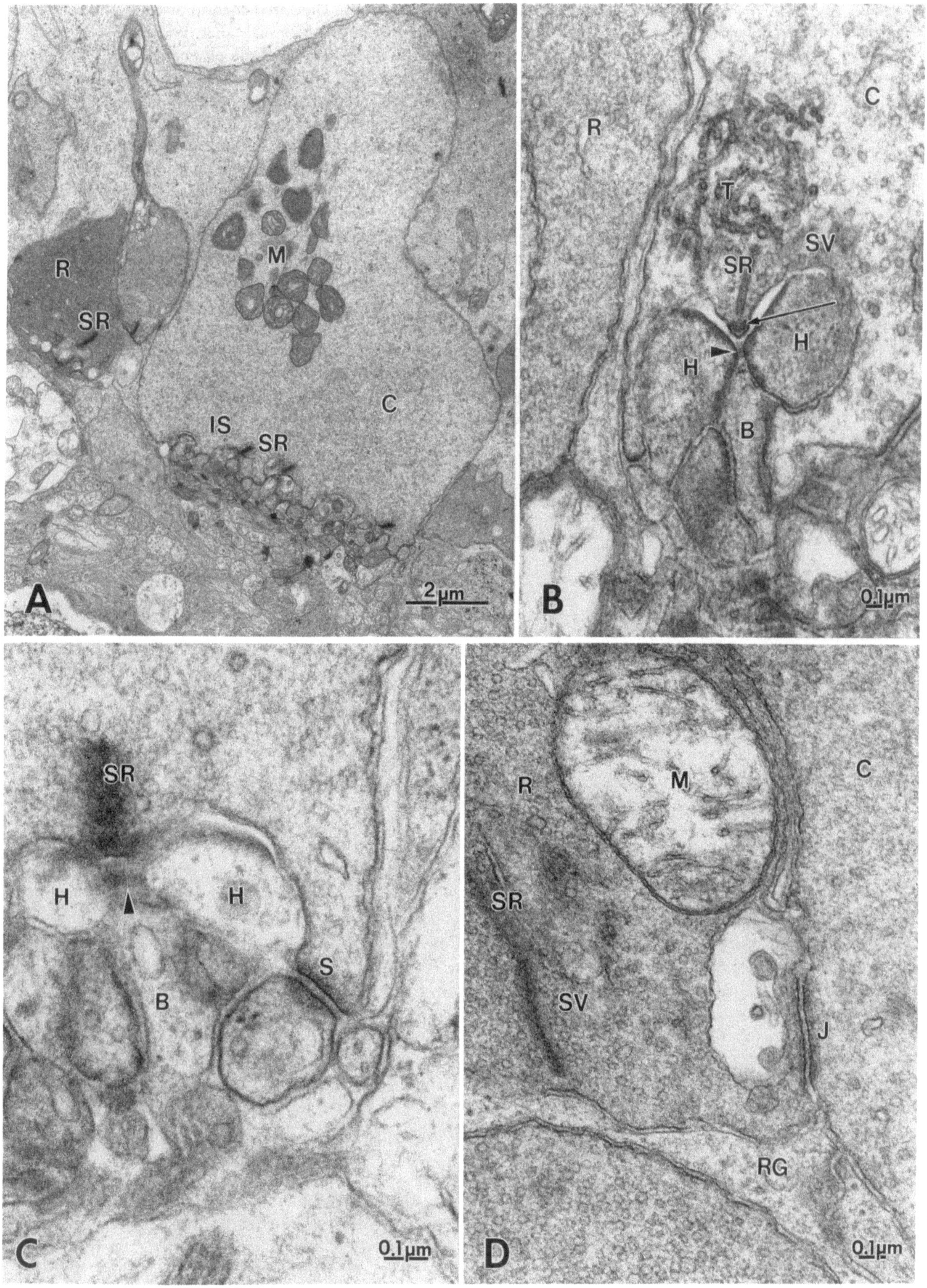
A
R
SR
M
C
IS
SR
2μm
B
R
C
T
SV
SR
H
H
B
0.1μm
C
SR
H
H
B
S
0.1μm
D
R
M
C
SR
SV
J
RG
0.1μm

of vertebrate photoreceptor cells. They are falciform plates in the presynaptic cytoplasm of the photoreceptor terminals that define the curvature of the dome shaped synaptic invaginations of the plasmalemma in the photoreceptor terminal. The synaptic ribbon is in the center of a ridge-like cytoplasmic projection into the cavity of the invagination (Fig. 24B). This synaptic ridge divides the synaptic invagination into two parts, semicircular in cross section. The synaptic ribbon stains well with the usual electron microscopic stains and seems to consist mainly of protein (Bunt 1971). It has been suggested that it is a "conduit" for synaptic vesicles to the presynaptic membrane (Dowling 1987, p. 48). Between the ribbon and the plasmalemma of the apex of the synaptic ridge is a trough-like accumulation of electron dense material (arrow). In cross section, its profile is shaped like the letter U, with the open end towards the ribbon. This material is termed the arciform density.

The contents of the invaginations in human rod and cone cells have been analyzed by Misotten (1965) in ultrathin serial sections. He found that in cone cells three postsynaptic processes penetrate into each invagination. This arrangement was called "triade." In the synaptic invagination of rod cells four endings were found and the term "tetrade" was coined. Later it was elaborated that the lateral processes within the invagination belonged to horizontal cells (H) and the central ones to bipolar cells (B) (Dowling 1987, pp. 53 ff). The lateral processes of horizontal cells in the invaginated synapses prevent the bipolar cell from coming into direct synaptic contact with the cytoplasmic membrane of the photoreceptor cell. Like a barrier between the bipolar cell dendrite and the synaptic ridge, electron-dense extracellular material (arrowhead) connects the postsynaptic horizontal cell terminals (H).

In Figure 24C a basal junction (S) is depicted. Such junctions occur in cone cell pedicles. They are not associated with synaptic vesicles; however, they are assumed to be chemical synapses (Dowling 1987, pp. 49 ff). In the specimens examined here, these junctions are located in the basal plasmalemma of the pedicles adjacent to invaginated synapses. The membranes in the junctional area are intensively stained, and their cytoplasmic surfaces are covered by electron-dense material. The junctional cleft contains stained material. It was shown that the postsynaptic partner of the basal junctions is a specific bipolar cell, the flat bipolar cell.

Similar junctions occur between rod cell spherules and cone cell pedicles (J). The nature of these contacts is not known. Gap junctions between photoreceptor terminals have been reported repeatedly (Dowling 1987, p. 52).

Figure 25 shows the mosaic of photoreceptor terminals in the peripheral retina of the baboon. This tangential section is from the same retinal column as the section through rod and cone inner segments represented in Figure 18A. The distribution of pedicles (C) and spherules (R) repeats the spatial pattern of cone and rod inner segments.

As the cone cell terminals are precisely aligned in parallel, all the pedicles in Figure 25 are cut through the same cytoplasmic domain that contains the mitochondrial cluster (M) at the transition from cone cell fiber to pedicle. Rod cells, on the other hand, are sectioned randomly through any level of the spherule. They do not terminate in a strictly defined plane as cone cells do. Their spherules are stacked in several rows, occupying the space between the cone cell pedicles.

The tangential section reveals that rod cell spherules also contain mitochondria (M). However, their number is small, usually two to four. The radial glial cells (RG) do not insulate the photoreceptor terminals from each other. There are many close appositions of the plasmalemma of both spherules and pedicles. Pedicles extend laterally cytoplasmic processes (asterisk) known as basal filaments (Polyak 1941, p. 248). The basal filaments contact other pedicles and spherules. A human cone may extend 6 to 12 basal filaments (Hogan et al. 1972, p. 456). As seen at the curved arrows in Figure 25, rod cell spherules possess basal filaments also. The function of these structures is not known.

Rod cell spherules contain only one invaginated synapse. Two synaptic ribbon profiles are seen in several invaginations (arrowhead) because the plane of the section cut through a single ribbon twice. Occasionally however, sperules have two synaptic ribbons (arrows).

Figure 25. Tangential section through layer of cone cell pedicles and rod cell spherules at level of mitochondrial cluster in pedicles; baboon, peripheral retina, ×3,800. C, cone cell pedicle; M, mitochondria; R, rod cell spherule; RG, radial glial cell; straight arrows, rod cell spherules with two synaptic ribbons; curved arrows, basal filament of rod cell spherule; arrowheads, single synaptic ribbons cut so that two separate sections are in plane of picture; asterisk, basal filament of cone cell pedicle.

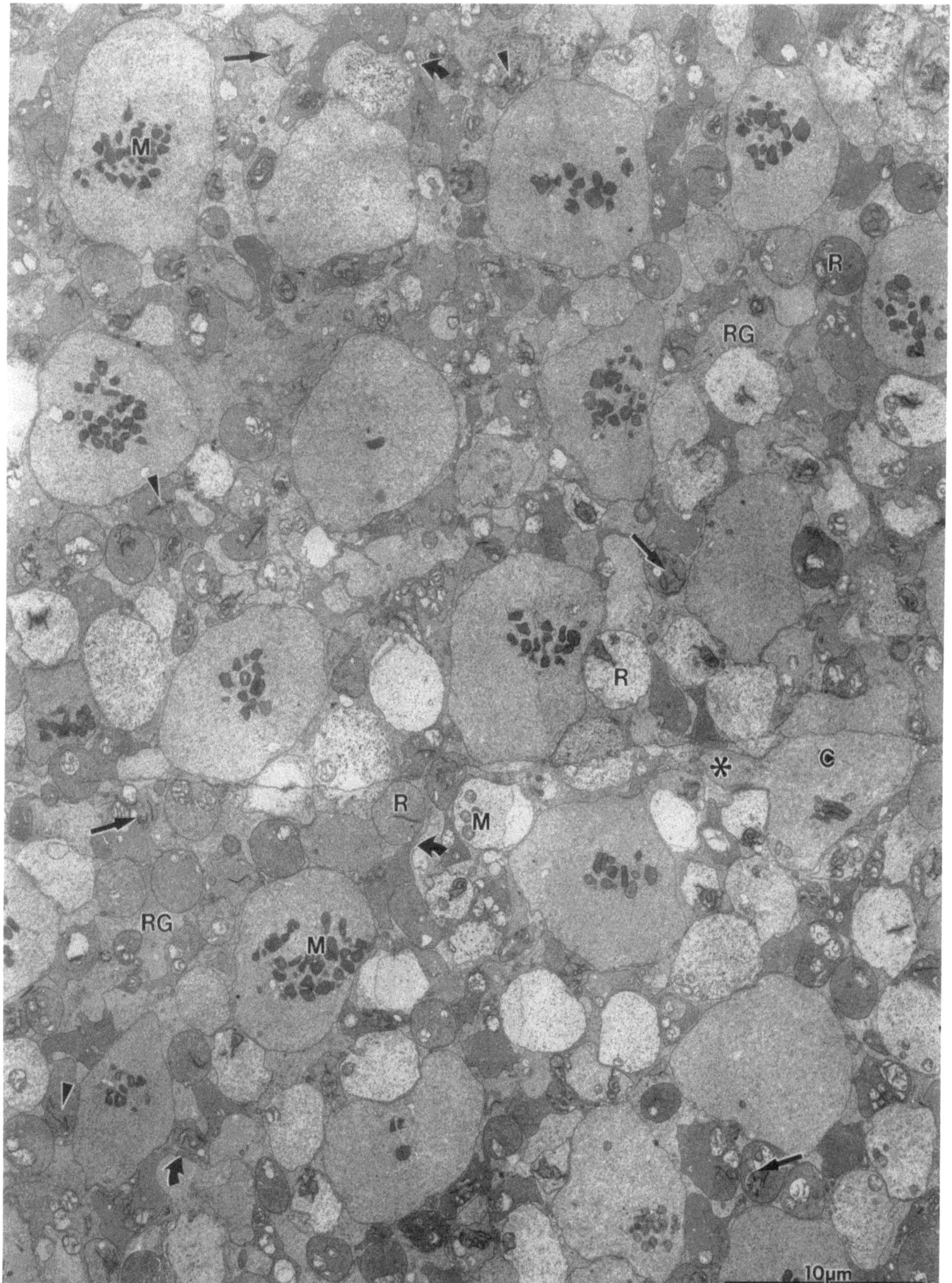
M
R
RG
R
C
R
M
RG
M
10µm

In **Figure 26** are seen tangential sections through the cone cell pedicles of the fovea. The locus from which these micrographs are taken is 500 μm from the foveal center (cross in Figure 7). The human cone cell pedicles are sectioned through the domain containing the cluster of mitochondria (M) (Fig. 26A). An occasional rod cell spherule (R) is seen. The pedicles are aligned loosely in parallel rows, separated from each other by the cytoplasm of radial glial cells. Adjacent pedicles within the rows come into direct contact frequently.

Figure 26B is a tangential section through the level of invaginated synapses. It is taken from the baboon fovea of which the cone inner segments are depicted in Figure 20. Thus, it may be that the pedicles seen here and the inner segments shown there belong to the same cone cells. The arrangement of the pedicles in parallel rows is also apparent. However, contacts between adjacent pedicles are less obvious than in Fig. 26A. Some punctate junctions are seen (arrowhead). Such junctions have been described as kissing junctions (Dowling 1987, p. 42). They are believed to be gap junctions.

Eleven to 15 synaptic ribbons (arrow) can be counted in primate cone cell pedicles. In the tangential section (Fig. 26B), often the synaptic ridge (arrow) is in the plane of the micrograph. The profile of the ribbon is seen in the center of the ridge, and vesicular structures are seen at its sides. The vesicular structures contain the postsynaptic terminals of horizontal cells, the lateral processes of the triade.

Figure 26. Mosaic of cone cell pedicles in fovea 500 μm from foveal center, ×3,800. (A) Tangential section through central part of pedicles, human. (B) Tangential section at level of invaginated synapses, baboon. C, cone cell pedicle; M, mitochondria; R, rod cell spherule; RG, radial glial cell cytoplasm; T, synaptic tubules; arrow, synaptic ridge; arrowhead, kissing junction between adjacent pedicles.

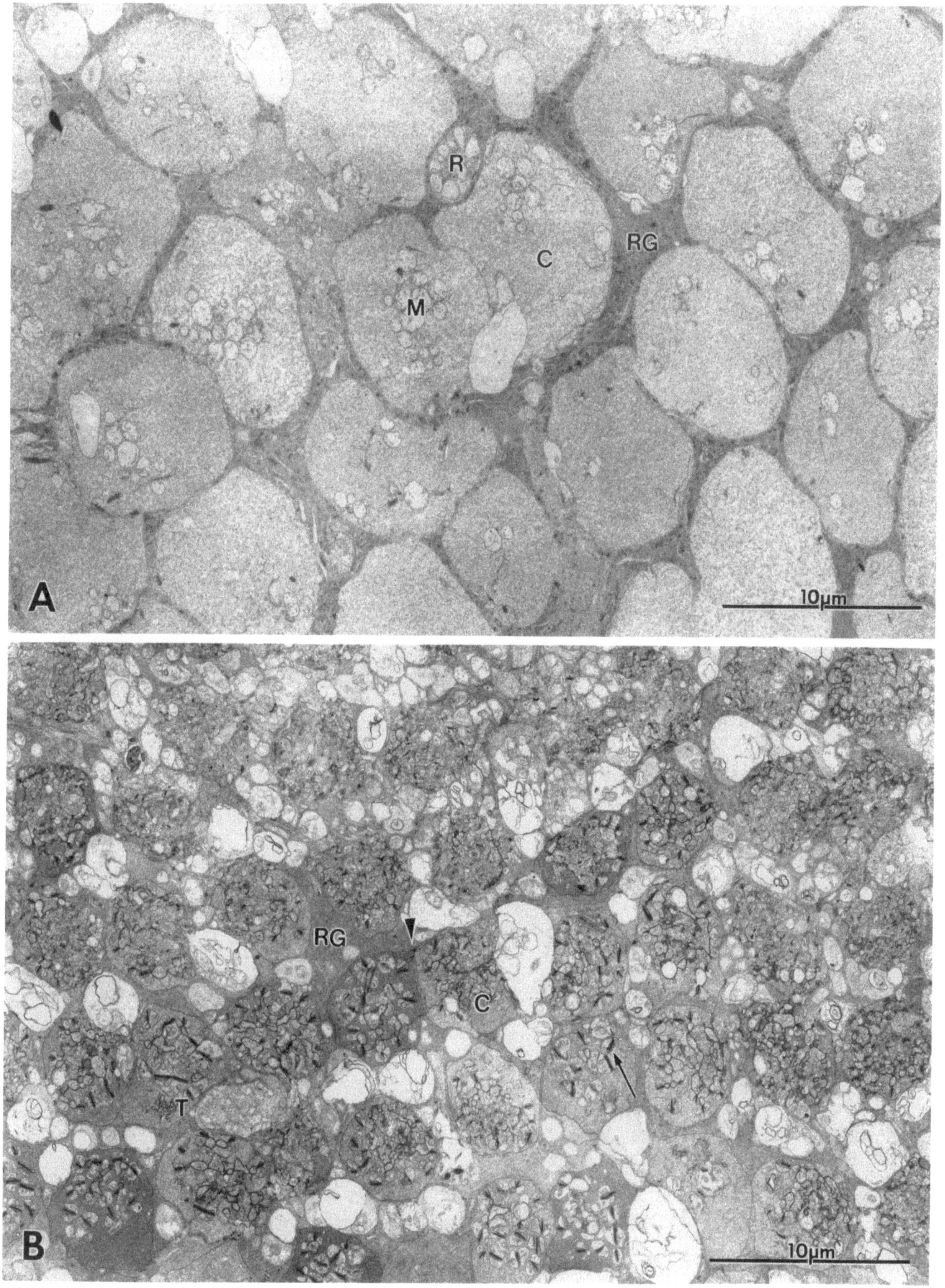
R
RG
C
M
A
10μm
RG
C
T
B
10μm

Figure 27 is a radial section that depicts the inner part of the foveal center of the baboon retina, directly underneath the central bouquet of cones. The major part of this retinal domain is filled by the electron-dense, featureless cytoplasm of radial glial cells. Two nuclei of these cells are seen (RG). There are always a few perikarya of neurons (B) in the foveal center (e.g., Polyak 1941, Figs. 38 and 39). Two neurons shown here presumably are bipolar cells. Supporting this assumption is the upper right neuron in Figure 27. This cell extends a process that contacts a cone cell pedicle (P) as the central element of the triade. Further, the chromatin pattern of the upper two nuclei (B) of the central foveal neurons resembles that of bipolar cells (see Fig. 33 for comparison). The perikaryon (H) in the lower left of Figure 27 may represent a horizontal cell because its chromatin is dispersed evenly in the karyoplasma (see Figs. 28 and 32 for comparison). The occurrence of pedicles in the central fovea has not been previously reported. In the fovea of which Figure 27 is taken, 19 displaced pedicles have been seen in its center in random sections. Horizontal cells in the center of the macaque fovea have been demonstrated (Röhrenbeck et al. 1990). Thus, the lateral postsynaptic processes in the invaginated synapses of displaced pedicles may derive from these horizontal cells. The displaced cone cell pedicles and their secondary neurons may not have significance for vision because their number is so low. However, they seem to be functional. In the upper right of Figure 27 a few cone cell fibers ("Henle's fibers") are seen.

Figure 27. Radial section through foveal center of baboon retina, ×8,000. B, bipolar cell perikaryon; CF, cone cell fiber ("Henle's" fiber); H, horizontal cell perikaryon; P, cone cell pedicle; RG, radial glial cell nucleus; thick arrow, center of foveal depression.

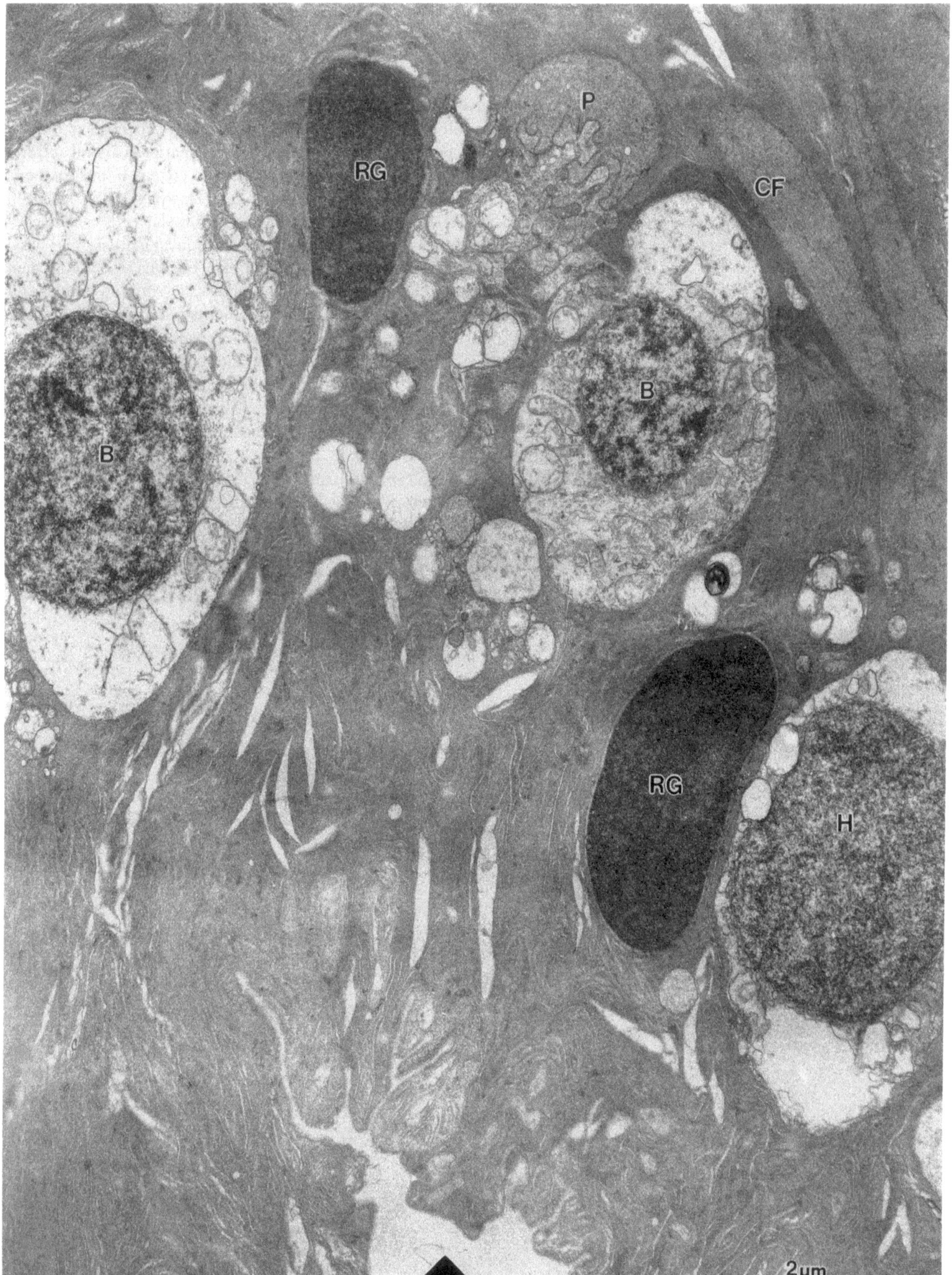
RG
P
CF
B
B
RG
H
2μm

Outer Plexiform Layer and Horizontal Cells

The outer plexiform layer contains fibers of horizontal and bipolar cells embedded in a matrix of radial glial cell extensions. In addition, some of the nervous profiles seen in this layer may derive from the interplexiform cells. The depth of the primate external plexiform layer is 3 to 4 μm.

The exterior margin of the external plexiform layer is defined by the flat bases of the cone cell pedicles. As discussed previously, processes of bipolar and horizontal cells penetrate into the synaptic invaginations of cone and rod cell terminals. The central postsynaptic processes in cone cell invaginations derive from a specific kind of midget bipolar cell (Dowling 1987, pp. 53 ff; Kolb and Lipetz 1991). Another variety of midget bipolar cells makes conventional contacts to cone cell pedicles close to the invagination. The basal junctions in cone cell pedicles connect to flat bipolar cells. Rod cell spherules are contacted by only one kind of bipolar cell. Any rod bipolar cell contacts several spherules, but the central elements in a given invaginated synapse of a spherule derive from different individuals of rod bipolar cells.

Primate retinae contain two kinds of horizontal cells (Gallego 1986), designated as type I and type II. They belong into the general group of horizontal cells with short axons. Axonless horizontal cells, which occur in most vertebrate species including several mammals, have not been found in primates.

The dendrites of both types of primate horizontal cells contact cone cells exclusively. All cone cell pedicles overlying the dendritic spread of these cells are contacted, and each pedicle is in synaptic contact with three to five horizontal cells of each type (Wässle et al. 1990).

Type I and II horizontal cells differ in their axonal arborizations. Axons of type I horizontal cells contact rod spherules exclusively, and those of type II cells terminate at cone pedicles. The axonal terminals penetrate into the synaptic invaginations and become lateral members of the postsynaptic complexes in spherules and pedicles, respectively. Axons of type I horizontal cells were found to be 800 to 2,500 μm long; type II cell axons travel a distance of 400 to 500 μm. It is believed that the axon terminal of horizontal cells is electrophysiologically independent from the soma and its dendritic extensions.

The pharmacology of the synapses made between photoreceptor, bipolar, and horizontal cells has been studied in lower vertebrates. According to these studies, l-glutamate is the transmitter released by the photoreceptor terminals, and gamma-aminobutyric acid (GABA) probably is the transmitter of type I horizontal cells. The transmitter of type II horizontal cells is not known. A detailed discussion of this topic is found in Dowling 1987 (pp. 133 ff).

It is not possible to identify the types of horizontal cells in the electron micrographs presented here. Likewise, the axon terminals could not be identified in the electron micrographs. Therefore, no reference will be made to the type of horizontal cells in the following.

Figure 28 is a radial section through the external plexiform layer of the peripheral baboon retina. A cone cell pedicle (C) is seen with three invaginated ribbon synapses. A synaptic ribbon is at SR and a basal junction at B. The pedicle makes junctional contact with an adjacent rod cell spherule (arrowhead), and another spherule contains a cluster of tubules (T) close to the synaptic ribbon. The postsynaptic processes of horizontal cells (H) are seen as the lateral endings in the synaptic invaginations of the pedicle. Many of the neuronal profiles in the external plexiform layer contain predominantly neurotubules. Such profiles probably are sections of axons (A) of horizontal cells. Dendritic processes (D) are seen also; they contain endoplasmic reticulum and other cytoplasmic organelles besides microtubules and mitochondria. The interstices between the neuronal profiles are filled with the cytoplasm of the retinal radial glial cells (RG).

The perikaryon of a horizontal cell dominates the lower part of Figure 28. Its cytoplasm is filled with numerous mitochondria (M), and a lysosome is seen at L. Rough endoplasmic reticulum (RER) and many free ribosomes stand out against the unstained ground substance. The nucleus (N) contains a distinct nucleolus. The chromatin is dispersed, giving the nucleus a homogeneously gray appearance.

At the arrow is a junctional contact between two dendrites. They may belong to horizontal cells, because they display the same electron-lucent cytoplasm as the horizontal cell described previously.

In horizontal cells of humans, but not in other primates, a peculiar cytoplasmic inclusion, Kolmer's crystalloid, has been described (Hogan et al. 1972, Figs. 9-52 and 9-53). This structure consists of an accumulation of straight parallel membraneous tubules, about 100 nm in diameter. The inclusion is approximately 0.4 μm in diameter. It stains with basic dyes and can be seen with the lightmicroscope. In the human retina used for this atlas, only one, not very typical example of Kolmer's crystalloid was seen.

Figure 28. **Radial section through external plexiform layer of peripheral retina of baboon, ×13,500. A, axon; B, basal junction; C, cone cell pedicle; D, dendritic process; H, horizontal cell process; L, lysosome; M, mitochondrion; N, nucleus of horizontal cell; NE, nucleolus; R, rod cell spherule; RER, rough endoplasmic reticulum; RG, radial glial cell; SR, synaptic ribbon; T, accumulation of tubules close to synaptic ribbon; arrow, junction between horizontal cell processes; arrowhead, punctate junction between rod cell spherule and cone cell pedicle.**

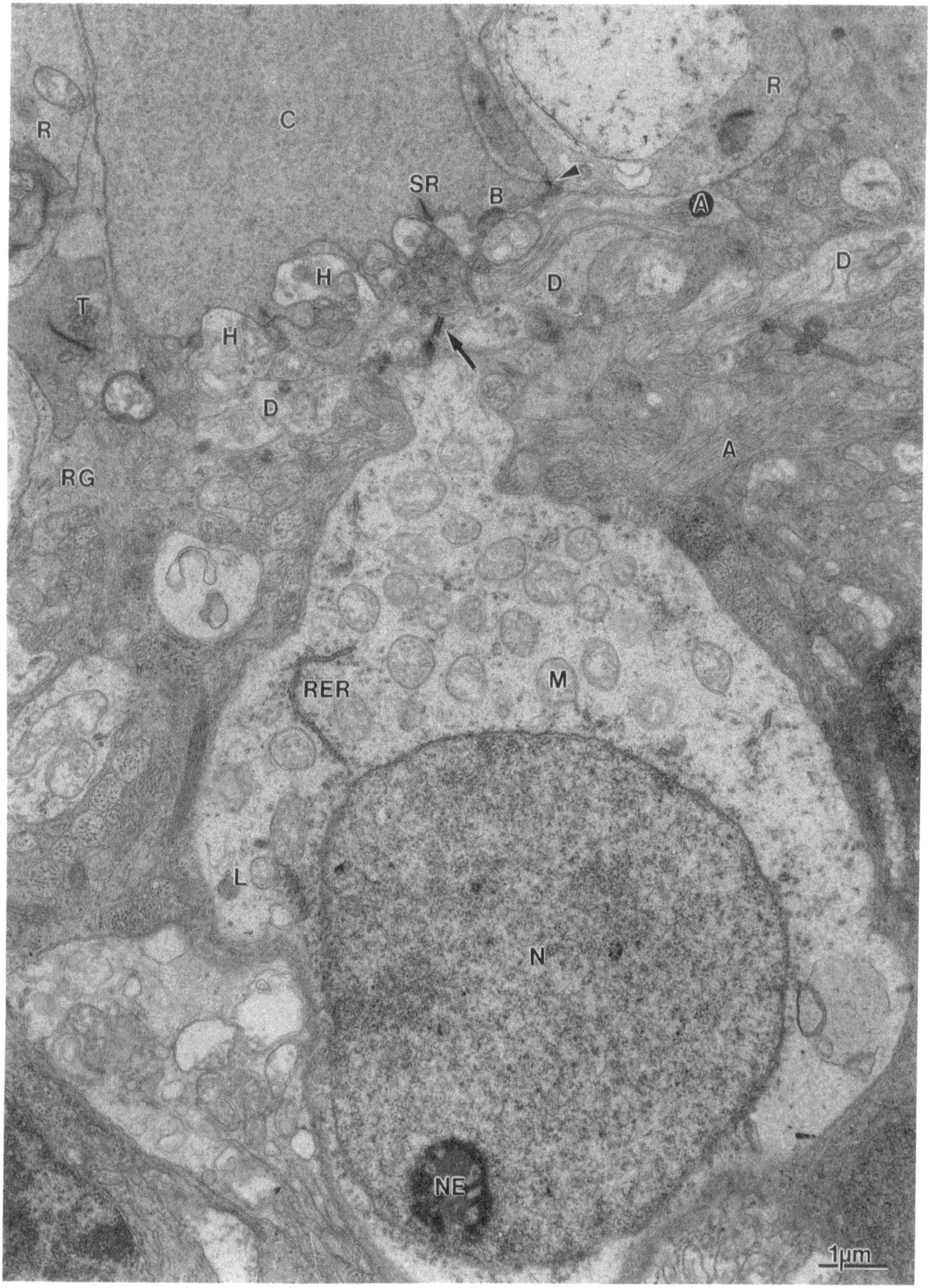
R
C
R
SR
B
A
H
D
D
T
H
D
A
RG
RER
M
L
N
NE
1μm

Figure 29 shows tangential sections through the external plexiform layer of cynomolgus monkeys. In the central part of Figure 29A, the bases of some cone cell pedicles are in the sectional plane. As this section is from the fovea, no rod cell spherules are present. The lateral parts of the micrograph cut through the center of the external plexiform layer. In these areas, at the asterisks, groups of fine nerve fibers are seen passing perpendicularly through the plane of the section just before they contact the pedicles. These short *fasciculi* are surrounded by tangentially oriented nerve fibers that run to various pedicles within the receptive fields of the horizontal or bipolar cells from which the fibers originate. In the lower left corner of Figure 29A, a perikaryon of a horizontal cell (H) protrudes into the external plexiform layer. Nearby, the wall of a capillary (C) is sectioned, indicating that the level of the capillary bed in the outer part of the external nuclear layer is immediately underneath the plane of this section.

Figure 29B depicts a subpedicular *fasciculus* of the peripheral retina. Nerve fibers with the characteristics of axons (A) and dendrites (D) are seen. They are attached to each other by numerous junctions. These junctions have been described by Hogan et al. (1972), p. 457) as desmosome-like. However, there are no intracytoplasmic filaments associated with them, nor is there the characteristic dark line in the junctional gap. The cytoplasmic membranes associated with the junctions are covered at their internal surfaces with regularly spaced particles; in the junctional cleft is electron-dense material (inset, Fig. 29B). Only some fibers contain clear vesicles (V) close to the junctions.

These features do not permit an unequivocal classification of the junctions as either synaptic or mere mechanical cell contacts. Assuming a desmosome-like nature of the junctions in accordance with the interpretation of Hogan et al. (1972), their possible role could be to tie together the nerve fibers destined to contact a particular cone cell pedicle. The fibers could also be sorted so that the correct cells are grouped together within the *fasciculus* to serve a particular synaptic invagination. However, should the junctions be synaptic, an additional pathway of information exchange may exist between horizontal and bipolar cells that is spatially separate from the cone cell pedicle.

Figure 29. Tangential sections through outer plexiform layer of cynomolgus monkeys. (A) 500 μm from foveal center, corresponding to location of cross in Fig. 7, ×3,900. (B) Peripheral retina, ×22,500; inset, ×66,000. A, axon; C, endothelial wall of capillary; D, dendrite; H, horizontal cell perikaryon; M, mitochondrion; P, basis of cone cell pedicle; RG, radial glial cell; V, vesicles; asterisk, neuronal extensions grouped together to contact cone cell pedicle; arrows, junctions.

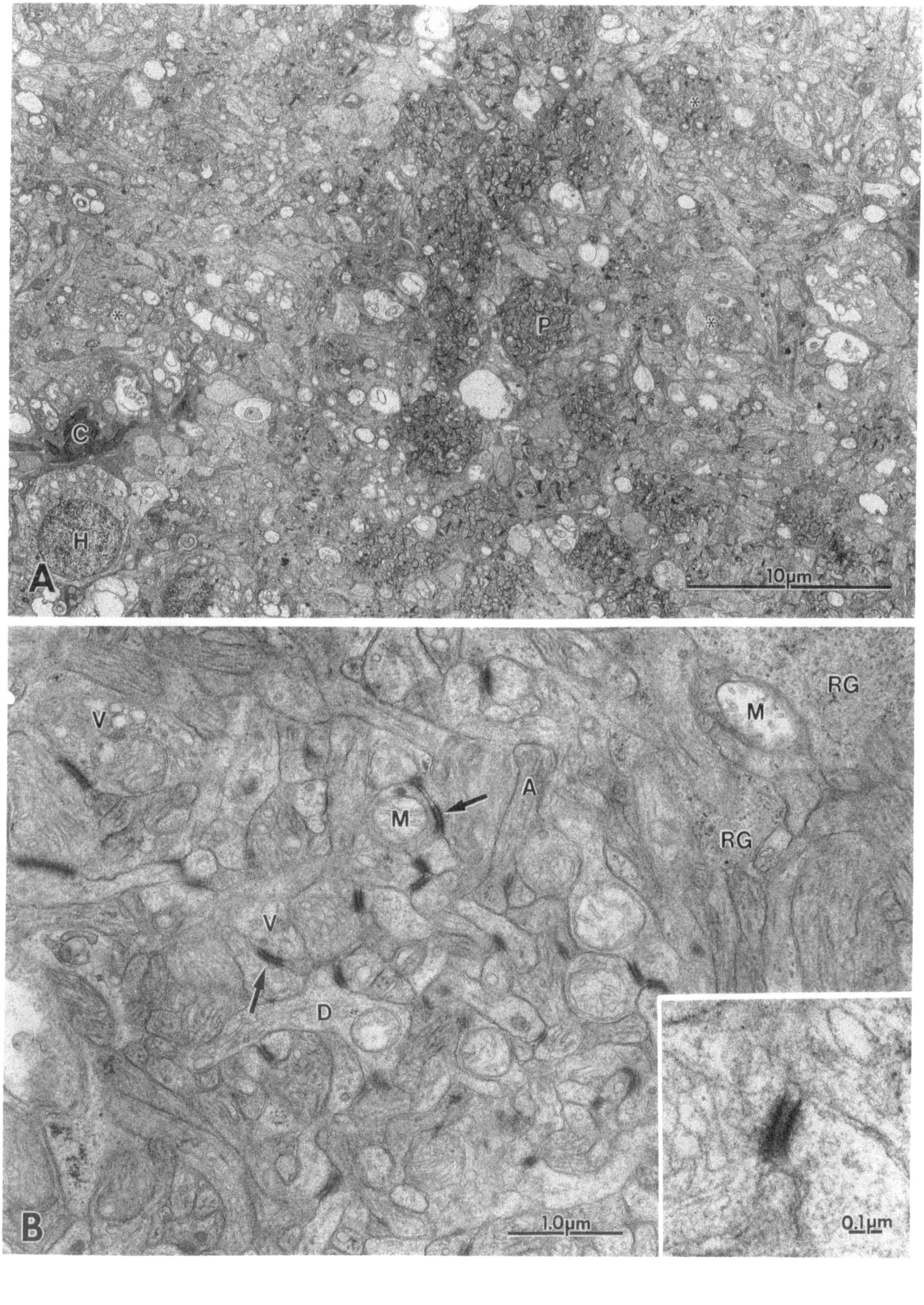
A
C
H
P
*
10µm
B
V
M
A
RG
D
1.0µm
0.1µm

Figure 30 shows horizontal cell bodies in a tangential section along the interface between the external plexiform and the internal nuclear layers. Figure 30A is from the peripheral retina of a cynomolgus monkey. The center-to-center spacing between the horizontal cell bodies (H) is approximately 13 μm. The space between them is filled with countless nerve fibers embedded in a matrix of the radial glial cells. Some of the fibers are cut perpendicularly to their axis. Most of these fibers derive from bipolar cells. Very few of them may belong to the interplexiform cells. Other fibers take their courses in the plane of the section (curved arrow). They may be bipolar cell dendrites, and axons or dendrites of horizontal cells.

In the parafoveal area of the baboon (Fig. 30B), the center-to-center distance between horizontal cell bodies (H) is approximately 10 μm. Here too, radial fibers (arrows) are sectioned that constitute dendrites of bipolar cells. Tangentially oriented nerve fibers are seen at the curved arrows. All interstices between the neuronal structures are filled with cytoplasm of the radial glial cell.

A blood capillary (C) is seen in the center of Figure 30B in a grazing section. The borders of the endothelial cells are seen as electron-dense lines (arrowheads). No space is detectable between the endothelial cells.

Figure 30. Interface between external plexiform and internal nuclear layer, ×3,900. (A) Peripheral retina of cynomolgus monkey. (B) Parafoveal retina of baboon. C, wall of blood capillary; H, cell body of horizontal cell; curved arrows, tangentially oriented nerve fiber; straight arrows, dendritic process of bipolar cells, oriented radially; arrowheads, border between adjacent endothelial cells.

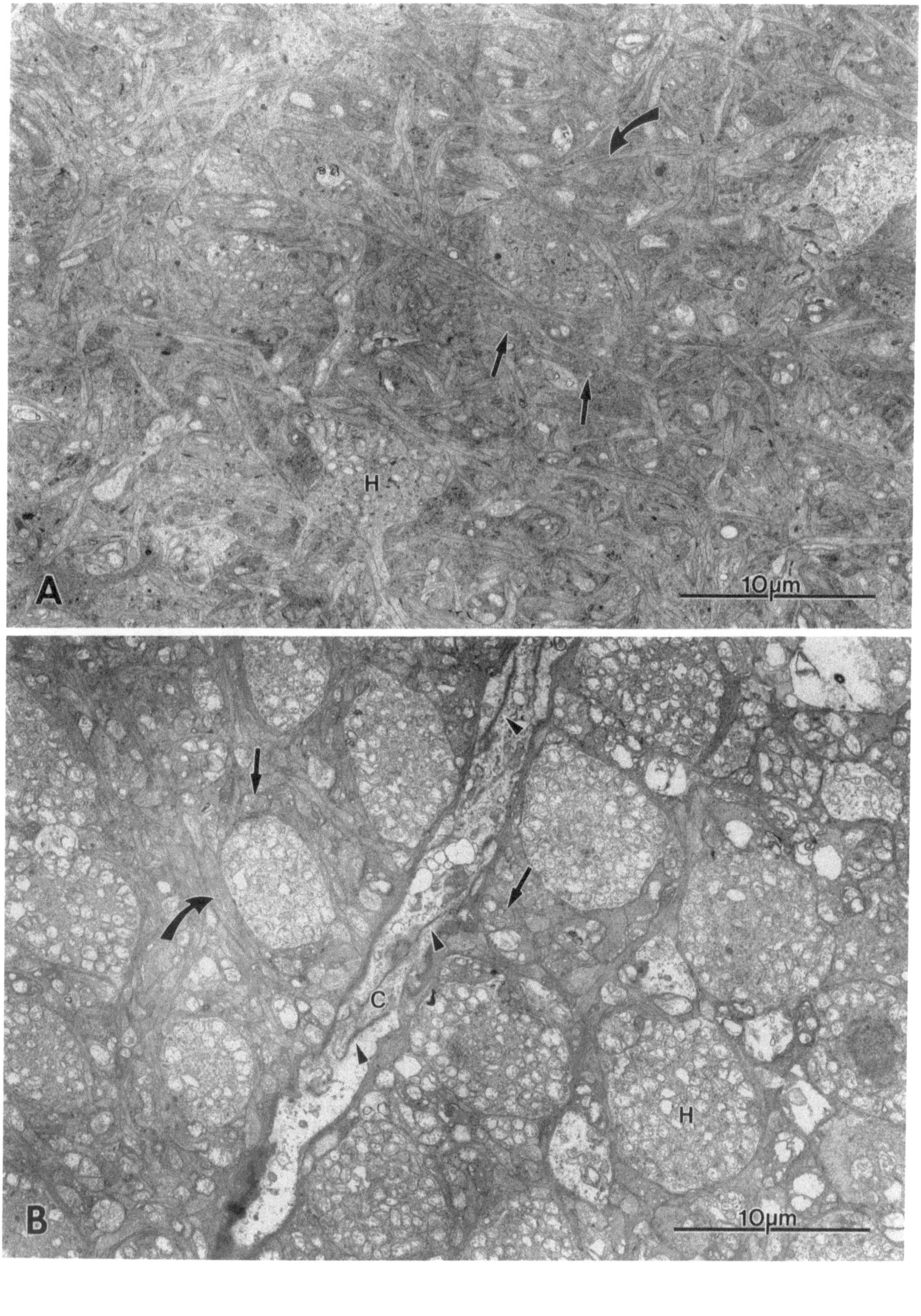

H
10µm
A
C
H
10µm
B

In the external part of the internal nuclear layer, shown in **Figure 31**, the horizontal cell perikarya form a monolayer. This level of the neuroretina is furnished with the external capillary bed. The wall of the capillaries (C) is of the continuous type. The endothelial cells are glued together by junctional complexes (arrowheads). A thin sheath of cytoplasm of the radial glial cells separates horizontal cell bodies from the wall of the blood capillaries (thick arrows). Erythrocytes (E) are seen in the lumen of the capillaries, because the retinae were fixed by immersion.

Dendrites of bipolar cells (straight thin arrows) and tangentially oriented nerve fibers (curved arrows) are seen between the horizontal cell bodies. The spaces between the neuronal and vascular structures are filled by cytoplasmic processes of the radial glial cells.

The horizontal cells of the fovea (Fig. 31A) are smaller than those in the periphery (Fig. 31B). The average center-to-center distance of the foveal horizontal cells is approximately 6.5 μm, in the periphery this distance is approximately 15 μm. In foveal horizontal cells, numerous small islets of electron-dense heterochromatin are randomly distributed in the nucleoplasm. Peripheral horizontal cells (Fig. 31B), on the other hand, exhibit pale nuclei without any heterochromatin aggregations. Foveal and peripheral horizontal cells have nuclei of approximately equal size. The ratio of nuclear to cytoplasmic volume of horizontal cells is larger in the fovea than it is in the periphery.

Figure 31. Tangential section through internal nuclear layer at level of horizontal cell nuclei, baboon, ×3,900. (A) Foveal sample, 500 μm from foveal center, location corresponding to cross in Fig. 7. (B) Peripheral sample. C, blood capillary; E, erythrocyte; H, perikaryon of horizontal cell; RG, radial glial cell cytoplasm; straight arrows, bipolar cell dendrites; curved arrow, tangentially oriented nerve fibers; thick arrow, sheath of radial glial cell cytoplasm between capillary wall and horizontal cell; arrowhead, border between adjacent endothelial cells.

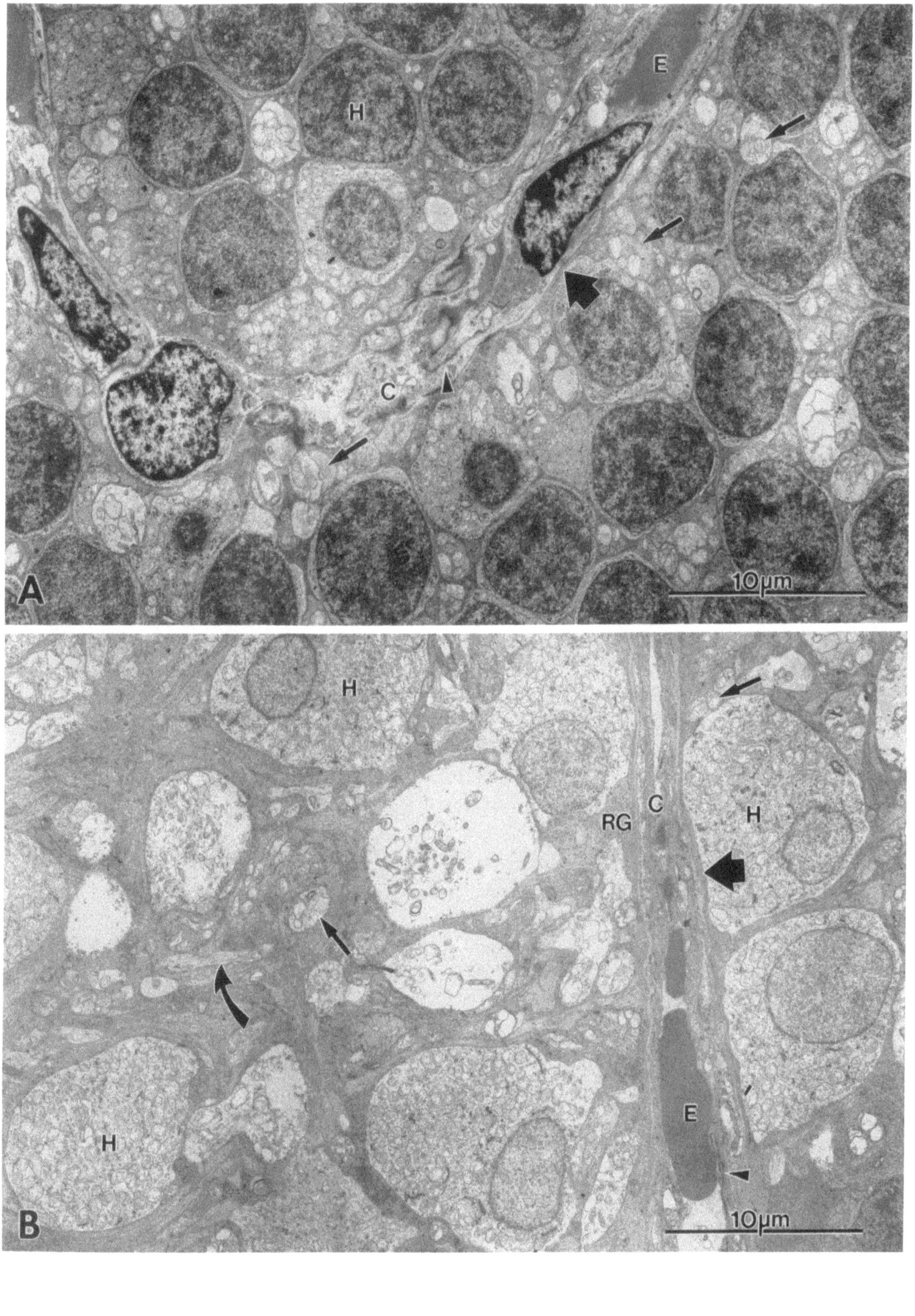
E
H
C
A
10μm
H
RG
C
H
H
E
B
10μm

Figure 32, which is from a tangential section of the peripheral baboon retina, shows details of horizontal cells. The cell seen in Figure 32A is cut eccentrically through the perikaryon. The pole of the nucleus (N) is in the plane of the section. The homogeneous distribution of chromatin is apparent. The cytoplasmic space is densely filled with organelles, such as rough (RER) and smooth (SER) endoplasmic reticulum, free ribosomes (R), mitochondria (M), lysosomes (L), and Golgi fields (G). All these features suggest that horizontal cells are highly active.

Thin sheaths (thick arrow) of radial glial cell cytoplasm (RG) separate the horizontal cell perikaryon from the basement lamina of a blood capillary (C). A process of a pericyte (PC) is in the upper right corner of Figure 32A.

In Figure 32B, a junction between a perikaryon and an axon of horizontal cells is depicted. These junctions are seen frequently in the external plexiform layer. They are characterized by electron-dense material in the junctional cleft and regularly spaced particles on the interior surface of the junctional plasmalemma. Occasionally small *cisternae* of the smooth endoplasmic reticulum (arrows) are associated with the junction. The nature of these cell contacts is not clear. They certainly do not resemble the communicating or gap junctions that have been demonstrated in horizontal cells of cold blooded vertebrates (Dowling 1987, pp. 50 ff). Typical communicating (gap) junctions between horizontal cells have not been found in the specimens described here.

Figure 32. Tangential section through horizontal cells of baboon peripheral retina. (A) ×14,000. (B) ×51,000. A, axon; BM, basement membrane; C, blood capillary; G, Golgi field; J, junction; L, lysosome; M, mitochondrion; N, nucleus; PC, pericyte; R, free ribosomes; RER, rough endoplasmic reticulum; RG, radial glial cell; SER, smooth endoplasmic reticulum; thin arrows, *cisternae* of smooth endoplasmic reticulum associated with junction; thick arrow, sheath of radial glial cell cytoplasm between capillary wall and horizontal cell.

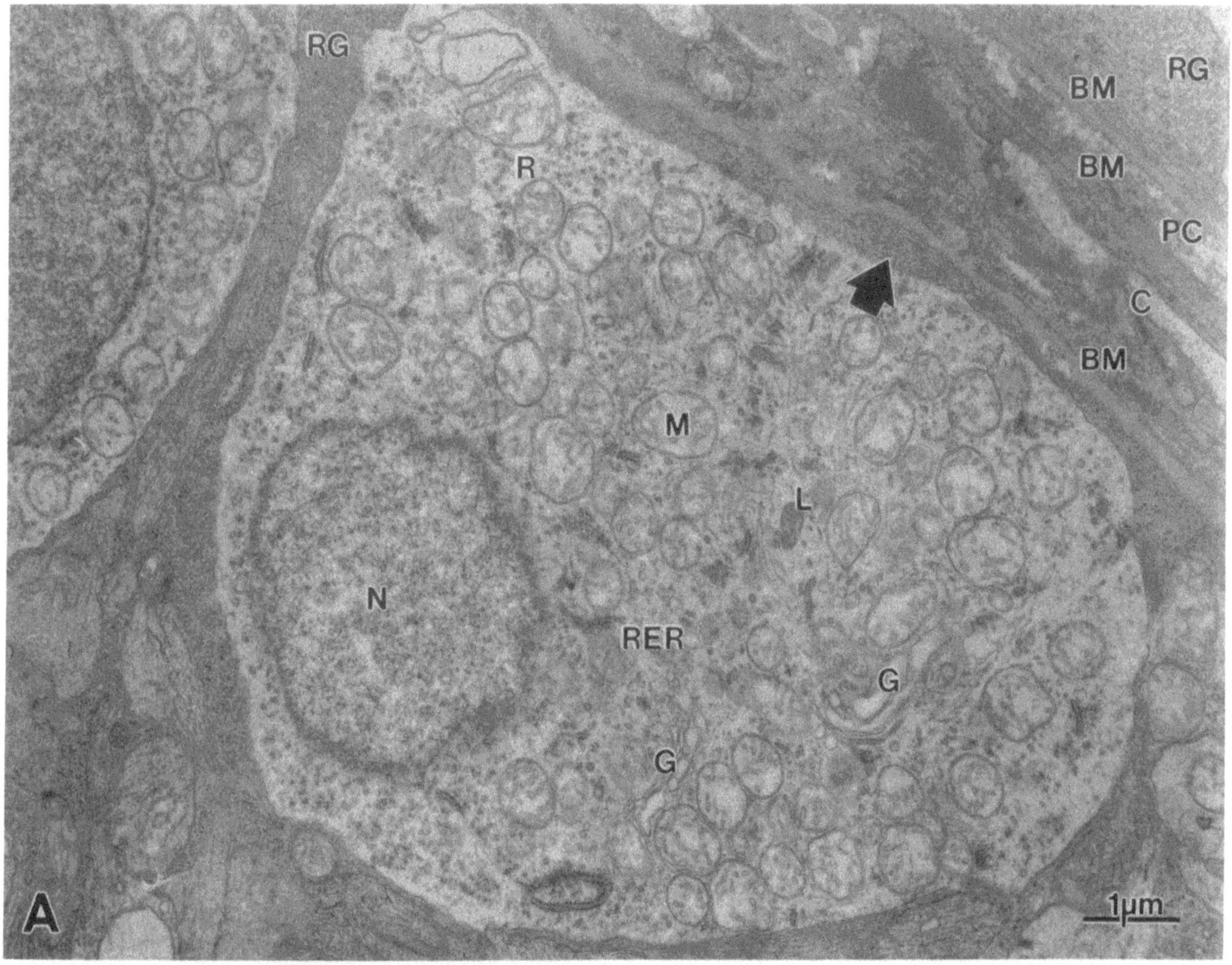
RG
RG
BM
BM
PC
C
BM
R
M
L
N
RER
G
G
A
1µm

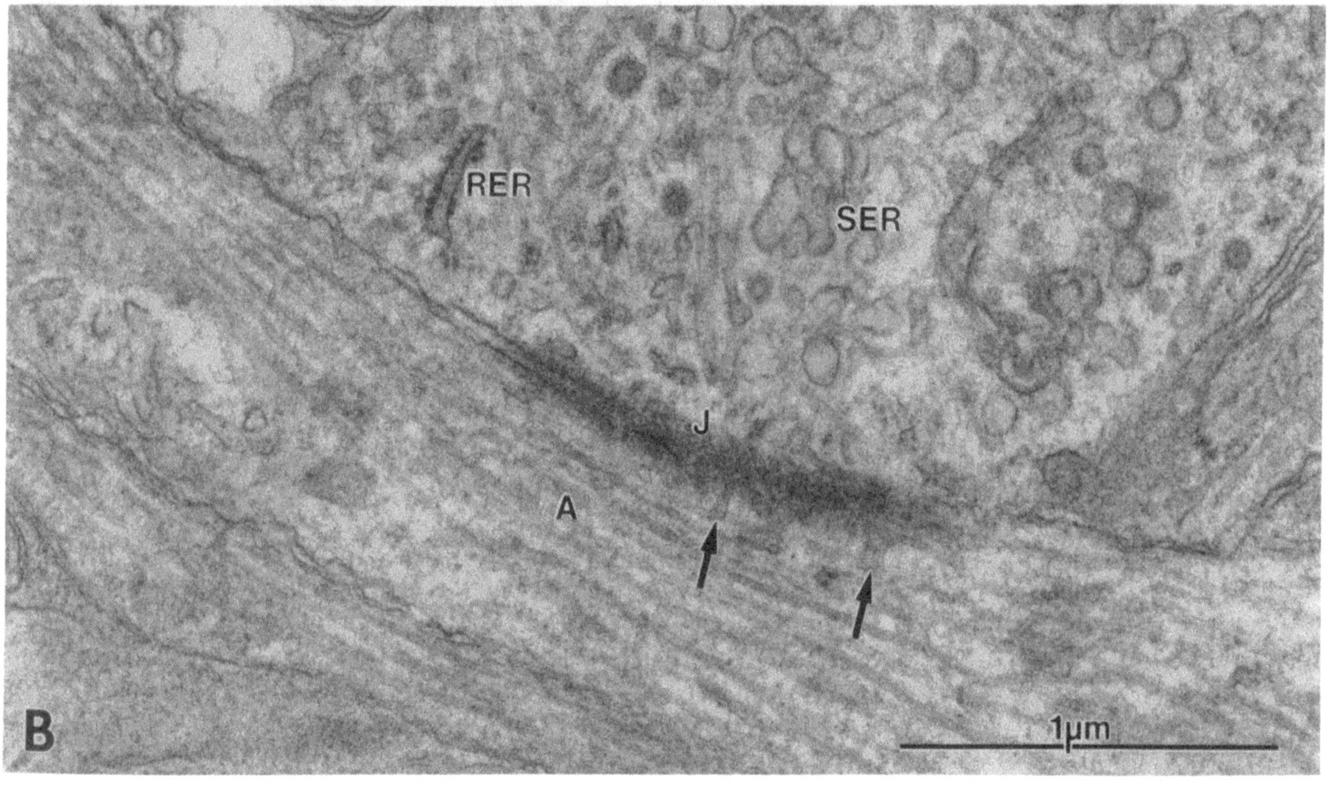
RER
SER
J
A
B
1µm

Bipolar, Radial Glial, and Amacrine Cells

Bipolar cells transmit information from the external to the internal plexiform layer. Cone and rod cells are connected to different sets of bipolar cells (Dowling 1987, pp. 25 ff). Rod cells connect to one type of bipolar cells, the rod bipolar cell.

Two major classes of bipolar cells are related to the cone system: flat bipolars, sending dendritic endings to the basal junctions of cone cell pedicles; and invaginating bipolar cells, contacting the pedicles via the central postsynaptic ending in the invaginated ribbon synapses. Flat and invaginated cone-related bipolar cells include midget bipolar cells. Any midget bipolar cell contacts a single cone cell, and any cone cell connects to one flat and to one invaginating midget bipolar cell. However, a nonmidget flat bipolar cell contacts several cone cell pedicles.

Invaginating cone bipolar cells become excited by an increase in light intensity that is centered on their receptive fields. These cells are called on-center bipolars. Their axons terminate in a distinct sublamina within the internal part of the internal plexiform layer, the on-sublamina. Flat bipolar cells, on the other hand, respond to decrease of the light intensity in their central receptive fields. They are termed off-center bipolar cells. Their axonal terminals spread in the external part of the internal plexiform layer, the off-sublamina (Kolb and Lipetz 1991).

Figure 33 is from a tangential section through the central part of the internal nuclear layer of the peripheral cynomolgus retina. The perikarya of bipolar cells predominate. Some bipolar cells are sectioned above or below the nucleus either through the axonic or dendritic fiber (arrows), or through the perikaryal cytoplasm (B) containing the majority of the cell's organelles (see Figs. 34 and 35). Other bipolar cells are cut through their nuclei (BN_1, BN_2, BN_3, BN_4).

Bipolar cell karyoplasm displays variable morphology. One type (BN_1) of bipolar cell nuclei exhibits fine granular chromatin with some small dense aggregations against a clear background. The second type (BN_2) is almost homogeneously gray because of the even distribution of finely granulated chromatin. Type three (BN_3) nuclei contain small dark islets of heterochromatin. The fourth type of bipolar cell nuclei (BN_4) is small, indented, and is filled with granularly dispersed chromatin. It is not possible at present to correlate the nuclear morphology with the subgroups of bipolar cells described earlier. The nucleoli (N) of bipolar cell nuclei are small.

Cytoplasm of the radial glial cells (RG) tightly fills the space between the bipolar cells. Some of their nuclei (RGN) are seen in the lower part of Figure 33.

Figure 33. Tangential section through central part of internal nuclear layer, cynomolgus, peripheral retina, ×3,800. B, perikaryal cytoplasm of bipolar cell; BN_1, BN_2, BN_3, BN_4, bipolar cell nuclei; C, blood capillary; RG, radial glial cell cytoplasm; RGN, radial glial cell nucleus; N, nucleolus; arrows, dendritic or axonal process of bipolar cell.

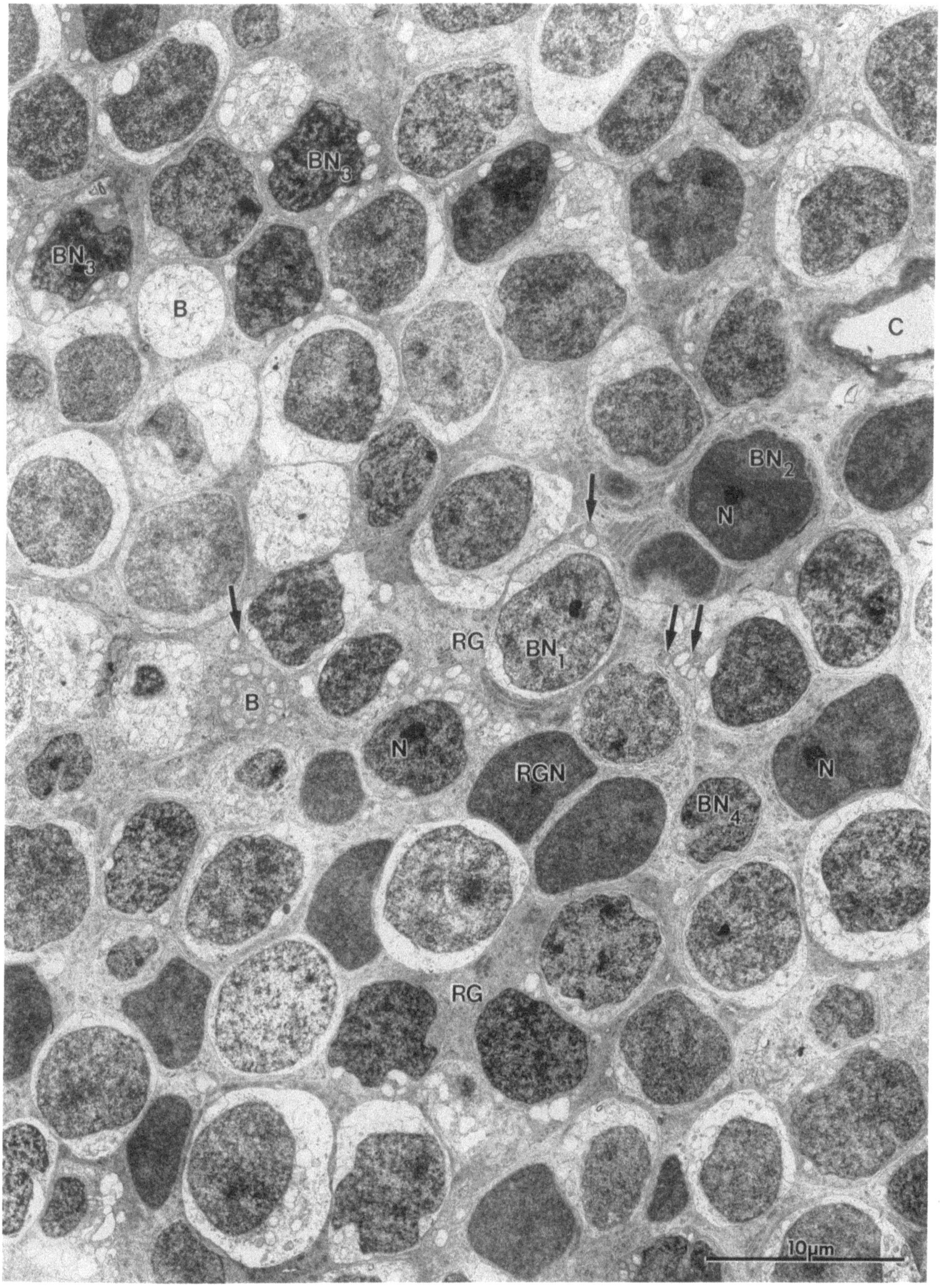
BN3
BN3
B
C
BN2
N
RG
BN1
B
N
RGN
BN4
N
RG
10μm

Details of bipolar cell fine structure are shown in **Figure 34**. In the upper left, the perikaryal cytoplasm of a bipolar cell (B) is shown containing several small Golgi fields (G), mitochondria (M), free ribosomes (R), lysosomes (L), coated vesicles (straight arrow), and subsurface *cisternae* of the endoplasmic reticulum (curved arrow). According to Misotten's analysis (Hogan et al. 1972, p. 461), the flat subsurface vesicles are characteristic constituents of rod bipolar cells, seen in their axons and perikarya.

In the center of Figure 34 is a perikaryon of a bipolar cell. It is 5.4 μm in diameter, and a coat of cytoplasm about 1 μm thick surrounds the nucleus. The nucleus contains granular, evenly distributed chromatin. Mitochondria (M), a centriole (C), a Golgi field (G), and microtubules (T) are seen in the cytoplasmic compartments surrounding the nuclei of bipolar cells.

Dendritic processes of bipolar cells (D) contain smooth endoplasmic reticulum (SER), mitochondria, and other organelles besides microtubules. The dendrite in the lower part of Figure 34 is 1.4 μm in diameter. The bipolar cells are completely covered with cytoplasm of the radial glial cell (RG). No contacts have been seen between bipolar cells within the internal nuclear layer.

Figure 34. Tangential section through center of inner nuclear layer of peripheral cynomolgus retina, ×20,400. B, bipolar cell; BN, bipolar cell nucleus; C, centriole; D, dendrite of bipolar cell; G, Golgi apparatus; L, lysosome; M, mitochondrion; R, free ribosomes; RER, rough endoplasmic reticulum; RG, radial glial cell; SER, smooth endoplasmic reticulum; T, microtubules; straight arrow, coated vesicle; curved arrow, subsurface *cisternae* of smooth endoplasmic reticulum.

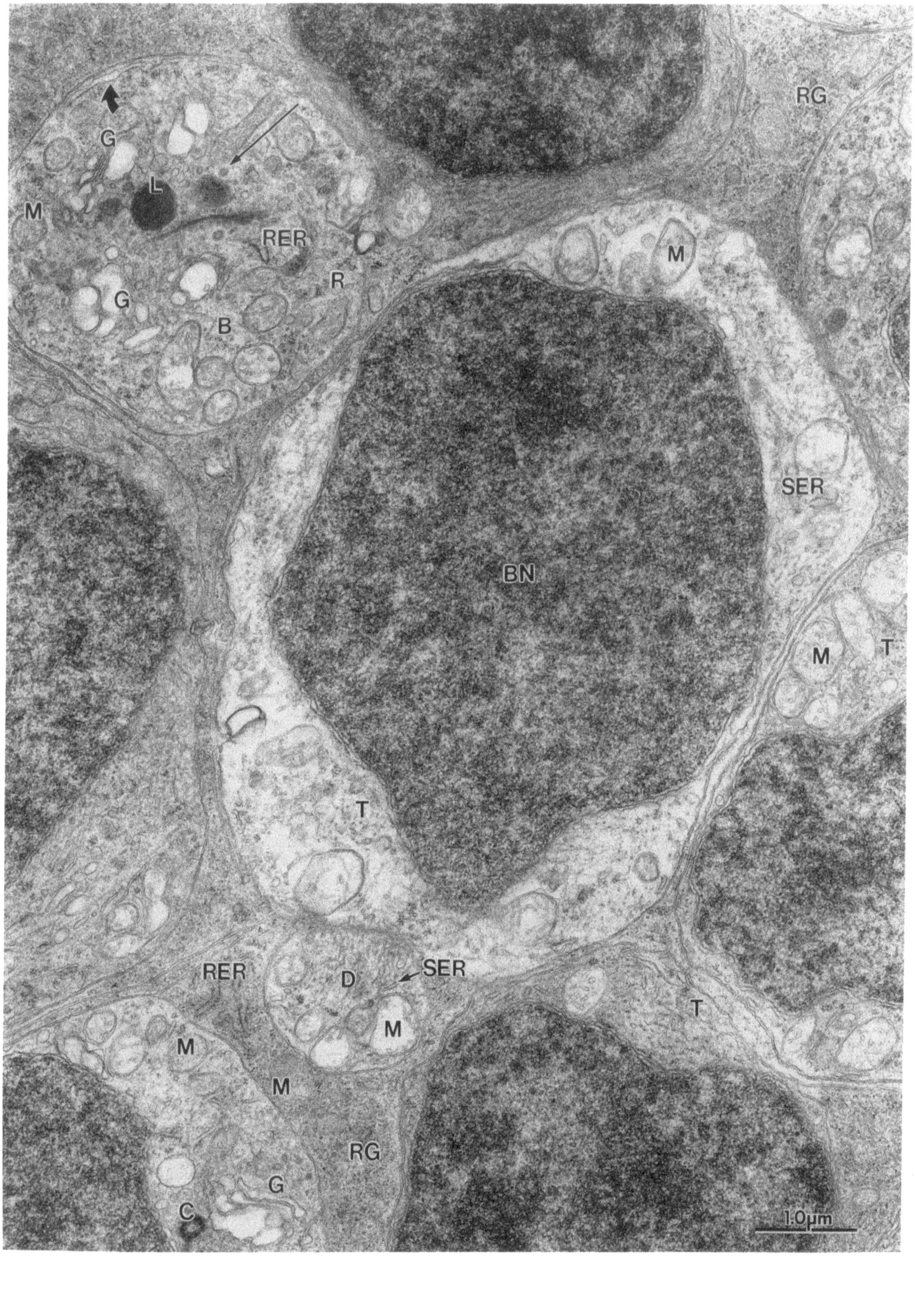
RG
G
L
M
RER
R
G
B
M
SER
BN
M
T
T
RER
D
SER
M
M
M
T
RG
G
C
1.0µm

Additional examples of the fine structure of bipolar cells are seen in **Figure 35**. Figure 35A depicts the perikaryal cytoplasm. Numerous mitochondria (M) are seen, containing tubular cristae. A Golgi field (G) exhibits coated vesicles; the centriole (C) is seen and smooth endoplasmic reticulum (SER) is sectioned. Free ribosomes are at R. The radial glial cell (RG) surrounding the bipolar cell contains a small Golgi apparatus (G).

Examples of dendrites (D) and axons (A) in cross section are demonstrated in Figure 35B. Axons contain as dominant cytoplasmic constituents microtubules and mitochondria (M). Flattened subsurface vesicles are seen in one axon (curved arrow), which may belong to a rod bipolar cell (see earlier).

In addition to microtubules and mitochondria, dendrites have rough and smooth endoplasmic reticulum and free ribosomes. One bipolar cell process in Figure 35B is filled completely by a mitochondrion (M). The nerve fibers seen in Figure 35B are between 0.4 and 0.8 μm in diameter.

Bipolar cell axons and dendrites are embedded in the cytoplasm of the radial glial cells (RG). The cytoplasmic membrane of the glial cells is invaginated around the nerve fibers similarly to mesaxons of Schwann cells in peripheral nerves (straight arrows). Some of the axons seen in tangential section of the center of the internal nuclear layer could belong to the interplexiform cells. However, as these cells are rare in comparison to

Figure 35. Tangential section through central part of inner nuclear layer, peripheral retina, cynomolgus monkey, ×20,000. (A) Perikaryal cytoplasm of bipolar cell. (B) Profiles of bipolar cell dendrites and axons. A, axon; BN, bipolar cell nucleus; C, centriole; D, dendrite; G, Golgi apparatus; M, mitochondrion; R, free ribosomes; RG, radial glial cell cytoplasm; SER, smooth endoplasmic reticulum; straight arrows, extracellular cleft between radial glial cells; curved arrow, subsurface cisterna in axon.

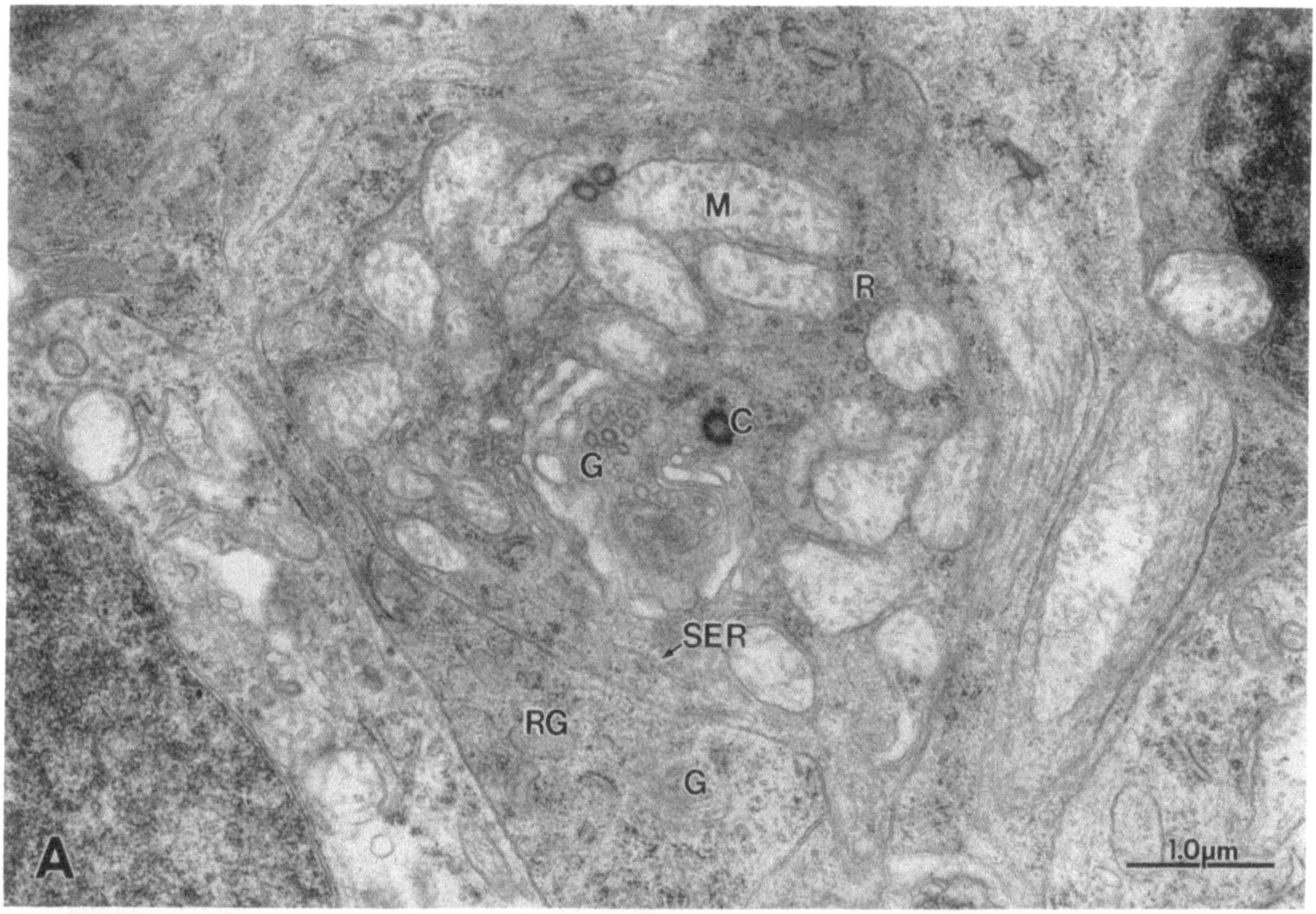
M
R
C
G
SER
RG
G
A
1.0μm

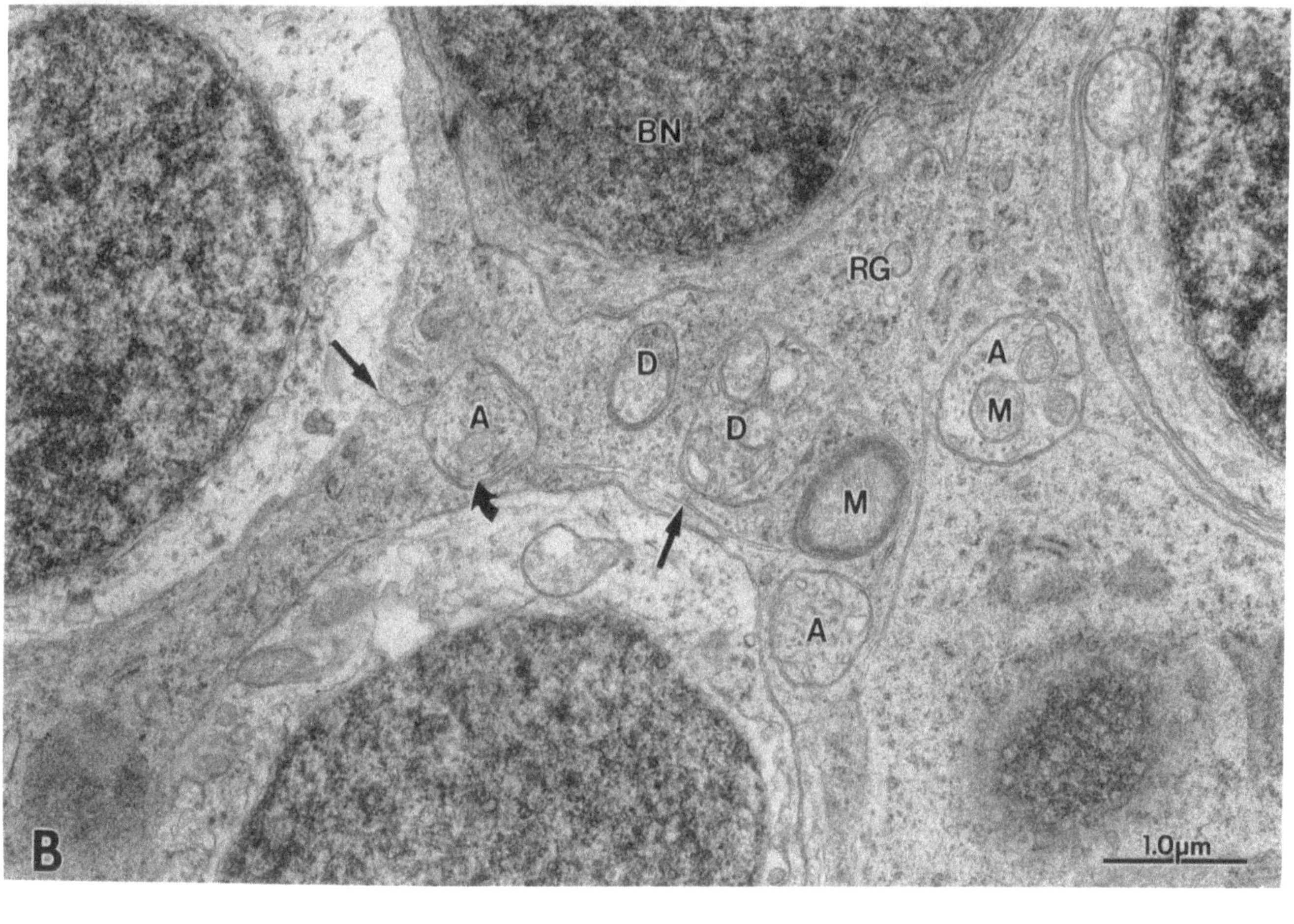
BN
RG
A
D
A
D
M
M
A
B
1.0μm

bipolar cells, the chance that one of the nerve profiles shown here belongs to them is slim. In electron micrographs, it is not possible to identify profiles of nerve fibers as originating from interplexiform cells.

Retinal radial glial or Müller cells are the principal glia of the retina, extending between the external and internal limiting membrane. Radial glial cell nuclei are situated in the central part of the internal nuclear layer. In radial sections of the retina, the band of glial cell nuclei is slightly undulating within the inner nuclear layer.

By light microscopy, the following domains can be distinguished in individual radial glial cells (Hogan et al. 1972, pp. 461 ff): *Radial processes*, extending from the perikaryon radially toward external and internal plexiform layers; *honeycomb meshwork*, cytoplasmic sheaths, visible with the light microscope around the perikarya of the retinal neurons in the outer and inner nuclear, and in the nerve cell layers; *horizontal fibers*, lateral cytoplasmic extensions between the nerve fibers of both plexiform layers; and *fiber baskets*, the light microscopical equivalent of the apical microvilli extending into the subretinal space.

Electron microscopy reveals that the spaces between the neuronal and vascular structures are tightly filled with the cytoplasm of the glial cells. In the outer nuclear layer, the photoreceptor cells, and in the inner nuclear layer, the bipolar cells, including perikarya and nerve fibers, are completely insulated from each other by radial glial cell cytoplasm (see Figs. 22, 23, and 33–35).

The function of retinal radial glial cells is as complex as their morphology. They are the major source of the b-wave of the electroretinogram, a slow, light-triggered voltage change with positive polarity toward the cornea (Rager 1979). A detailed discussion of this topic can be found in Dowling (1987, pp. 164 ff). Related to their electrical activity is the participation of the radial glia in driving the retinal flow of potassium ions (Reichenbach 1989). As do astrocytes in other parts of the central nervous system, retinal radial glial cells actively remove excess potassium ions, produced by the activity of the neurons, from the extracellular compartment. The vitreous is used by the glial cells as buffer volume to store or withdraw potassium ions.

The radial glial cells play a vital role in the turnover of substances excreted or taken up by the retinal neurons. One example is the metabolism of gamma-aminobutyric acid (GABA) and glutamatic acid (Riepe and Norenburg 1977). Radial glial cells take up GABA and glutamic acid, which are released as neurotransmitters by retinal neurons. In the glial cytoplasm these substances are converted into glutamin, which diffuses back into the neurons. GABA or glutamate are resynthesized in the neuronal cytoplasm to be reused as transmitter. Retinal radial glial cells store large amounts of glycogen.

Figure 36 is from a tangential section through the central part of the inner nuclear layer of cynomolgus monkey. At the area depicted, the undulating monolayer of retinal radial glial perikarya is in the plane of the section. The nuclei of the glial cells (RGN) contain homogeneously dispersed chromatin, resulting in an amorphous, dark gray appearance of their karyoplasm. Their nucleolus (N) is small and electron-dense. The radial glial nuclei are 5 to 6 μm in diameter.

Some perikarya of bipolar cells (B) are seen between the radial glial cells in Figure 36A. Bipolar cell dendrites and axons (arrows) are seen as small circular profiles embedded in the perikaryal cytoplasm of the radial glial cells.

Figure 36B shows some details of the cytoplasm in the perikaryon of the radial glia (RG). Regular stacks of the rough endoplasmic reticulum are characteristic for these cells. There are microtubules (T) cut perpendicularly, mitochondria (M), lysosomes (L), and free ribosomes (R). A dendrite (D) of a bipolar cell contains mitochondria (M) with tubular cristae.

Figure 36. Tangential section through level of perikarya of radial glia, cynomolgus monkey. (A) ×3,600; (B) ×17,600. B, bipolar cell; D, dendrite; L, lysosome; M, mitochondrion; N, nucleolus; R, free ribosomes; RER, rough endoplasmic reticulum; RG, radial glial cell cytoplasm; RGN, nucleus of radial glial cell; T, microtubules in cross section; arrows, dendrite or axon of bipolar cell.

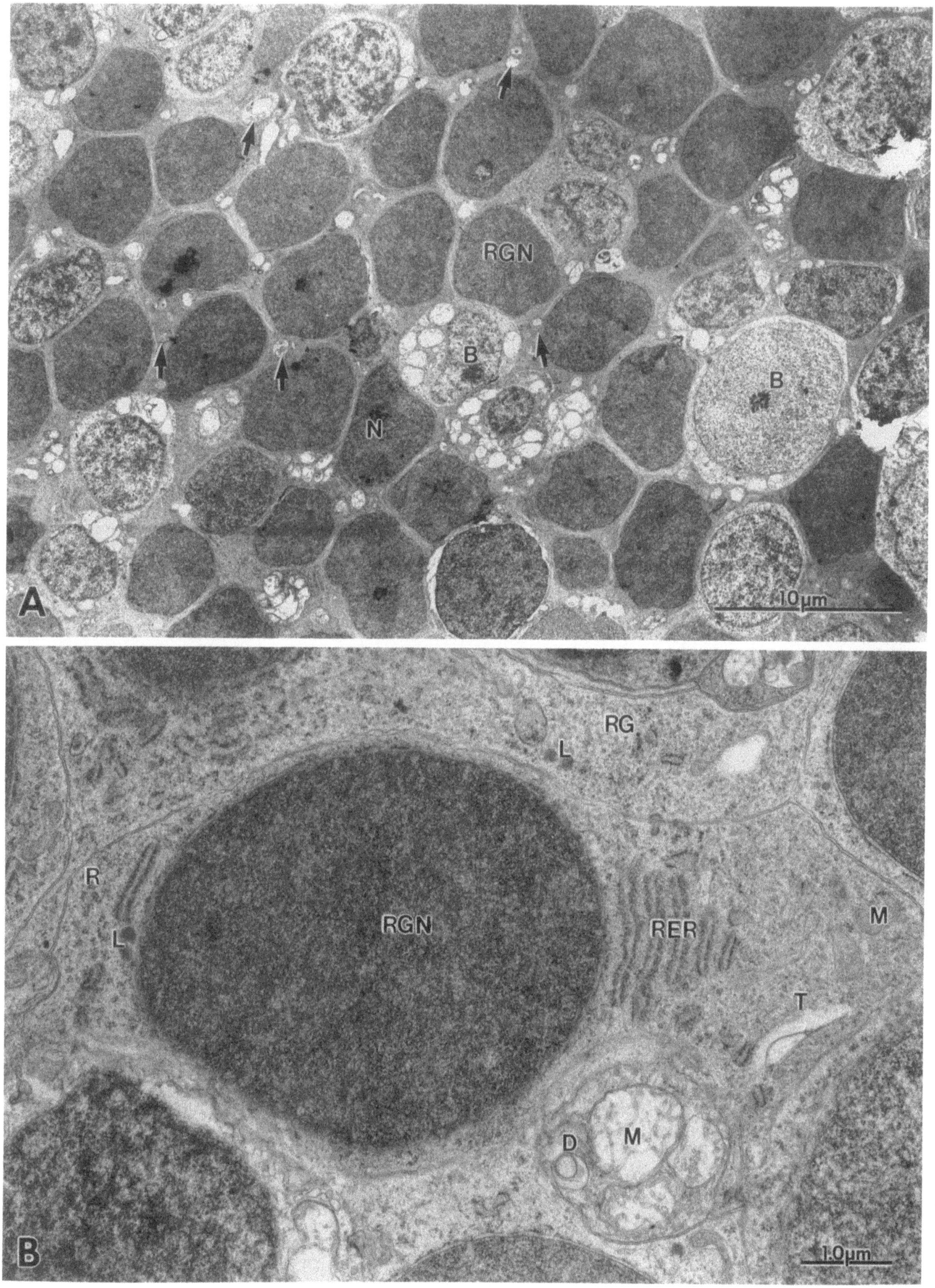
RGN
B
B
N
10µm
A
RG
L
R
L
RGN
RER
M
T
D
M
1.0µm
B

Another example of the perikaryal domain of radial glia is shown in **Figure 37**. The dark gray nucleus is at N. As also seen in the previous figure (Fig. 36), the overall density of the radial glial cell cytoplasm is relatively high. There are mitochondria (M) and rough endoplasmic reticulum (RER). The perikarya interdigitate intensively, and small junctions (arrows) are frequently seen between adjacent glial cells. One such junction is shown at higher magnification in the inset. It is characterized by a 20 nm wide junctional cleft. The cytoplasmic surface of the plasmalemma at the junction is covered by electron-dense material, which seems to consist of globular subunits. These junctions resemble *maculae adherentes*.

A bipolar cell perikaryon (B) is sectioned above or below the nucleus. It contains mitochondria, rough endoplasmic reticulum, and free ribosomes. The cytoplasmic membranes of the bipolar cell and of the glial cell surrounding this profile are cut at a flat angle. Thus, the membranes are not visible as defined lines; they are seen merely as a gray area of uncertain thickness (asterisk). Another profile of a bipolar cell perikaryon contains a nucleus (BN). A bipolar cell axon, seen in cross section at A, contains mitochondria and neurotubules.

Figure 37. Tangential section through perikaryon of retinal radial glial cell, peripheral retina of cynomolgus monkey, ×21,000; inset, ×54,000. A, axon; B, bipolar cell cytoplasm; BN, bipolar cell nucleus; M, mitochondrion; N, nucleus of retinal radial glial cell; R, free ribosomes; RER, rough endoplasmic reticulum; arrows, desmosome-like junction; asterisk, cytoplasmic membranes cut at flat angle.

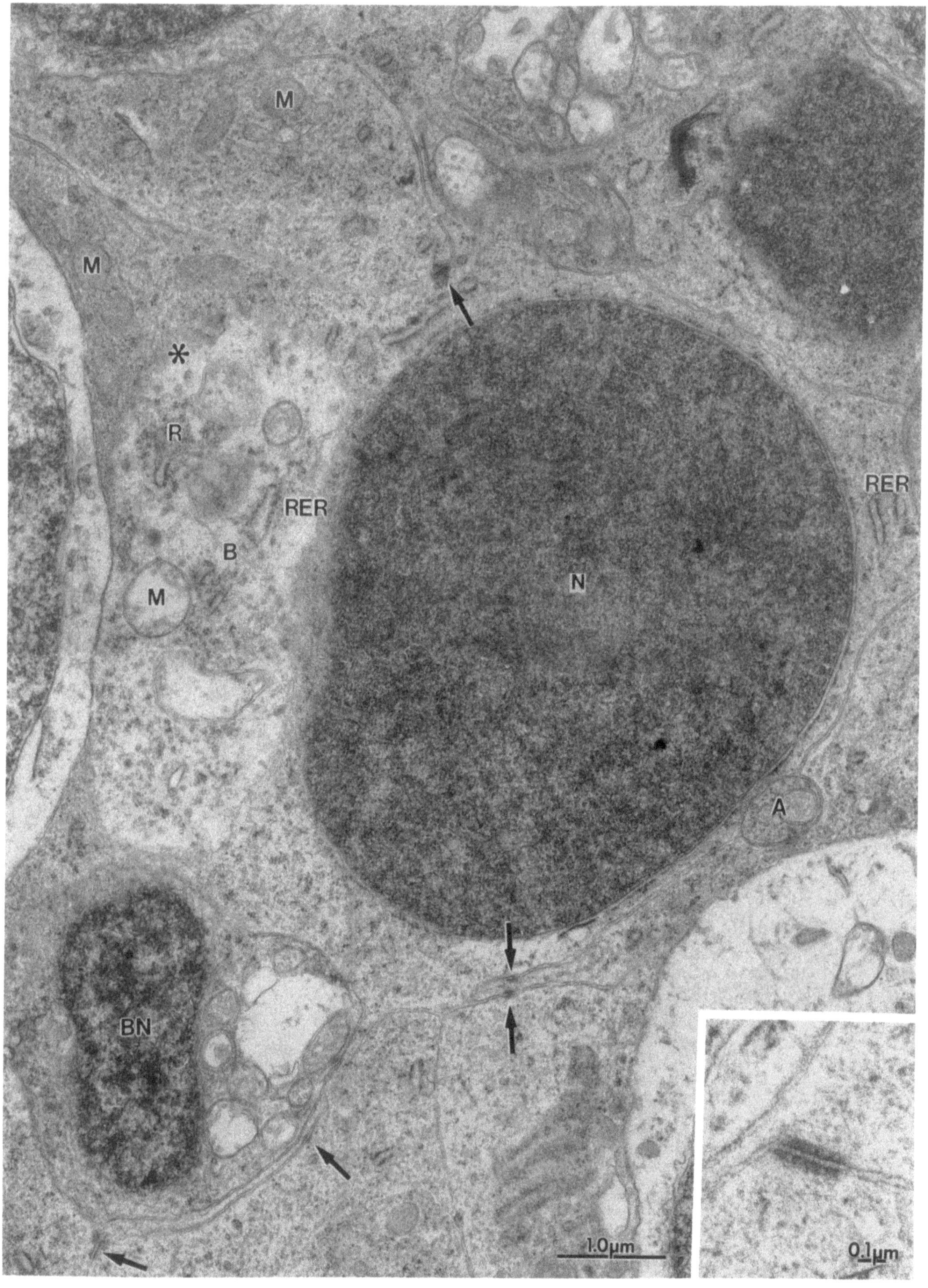
M
M
*
R
RER
B
M
RER
N
A
BN
1.0µm
0.1µm

The term amacrine cell includes a highly diverse group of cells with the common characteristic that their fibers are restricted to the inner plexiform layer. Masland (1988) has estimated the number of different kinds of amacrine cells to be at least ten, but possibly twenty or more. This diversity reflects differences in morphology as well as in connectivity and in pharmacology. Dowling (1987, p. 139) listed five neuroactive substances and eight neuropeptides found in amacrine cells.

Masland (1988) suggested a convenient scheme to group amacrine cells into three classes: The first class comprises amacrine cells that are densely packed and have small dendritic fields. The amacrine cell known as AII belongs into this class. AII cells receive input from rod bipolar cells and connect to ganglion cells; rod bipolar cells do not contact ganglion cells directly. Other narrow-field amacrine cells may be involved in the direct transmission of signals from cone specific bipolar to ganglion cells.

The second class contains sparse, widely branched amacrine cells. These cells have a low spatial density but their dendritic trees spread widely, sometimes as much as 1,000 μm. One example of this kind of amacrine cell is the dopaminergic amacrine cell. A single such cell affects the activity of approximately 41 retinal ganglion cells. Thus, amacrine cells of this class have a global regulatory function.

The third class of amacrine cells encompasses frequent cells with wide and overlapping dendritic fields. For example, the cholinergic amacrine cell belongs to this class. This cell occurs in large numbers, and its dendritic tree spreads for 250 to 800 μm. There is evidence that local domains in the dendritic field of a cell of this type have functional autonomy. Thus, information received locally does not spread over the whole cell. This mechanism would prevent loss of spatial resolution inherent in the information conveyed by these amacrine cells.

Polyack (1941, p. 298 ff) included a retinal neuron that he termed centrifugal bipolar cell into the class of amacrines. However, this cell type, now known as an interplexiform cell, has an ascending axon into the external plexiform layer. It should be considered a separate class. Details regarding morphology, function, and pharmacology of amacrine and interplexiform cells, and the role they play in vision, are discussed in Dowling's (1987) book.

In the inner nuclear layer, the somata of amacrine cells form a distinct sublayer adjacent to the internal plexiform layer (see Figs. 2 and 6). The correlation of amacrine cell ultrastructure with the amacrine cell types just described has not sufficiently been determined to allow an undubitable classification of the cellular profiles seen in the following electron micrographs.

Figure 38 depicts the layer of amacrine cell bodies in a tangential section of the cynomolgus fovea. The diversity of neuronal profiles in this section is apparent. There are large perikarya of amacrine cells containing a lobulated nucleus (A_1). Their cell bodies are approximately 10 μm in diameter, and their cytoplasm is electron-translucent. One of these large cells contains myelinoid bodies (MY) in its cytoplasm. The nucleus of these cells contains finely granulated chromatin and a distinct nucleolus (N).

A second type of amacrine cells (A_2) also has a lobulated nucleus, but its perikaryon is smaller, approximately 6 μm in diameter. The third kind of amacrine cell profiles (A_3) shows an unlobulated nucleus containing homogeneously dispersed chromatin. The cells marked as A_4 are the smallest amacrines, only 3 μm in diameter. Their nucleus is more electron-dense than that of the cells described above. Some cellular profiles are perikarya of bipolar (B) or of retinal radial glial cells (RG).

Figure 38. Tangential section through sublayer of internal nuclear layer containing amacrine cell perikarya, cynomolgus monkey, fovea, 500 μm from center (cross in Fig. 7), ×3,200. A_1, A_2, A_3, A_4, amacrine cell perikarya; B, bipolar cell perikaryon; MY, myelinoid body; N, nucleolus; RG, retinal radial glial cell; arrow, bipolar cell axon.

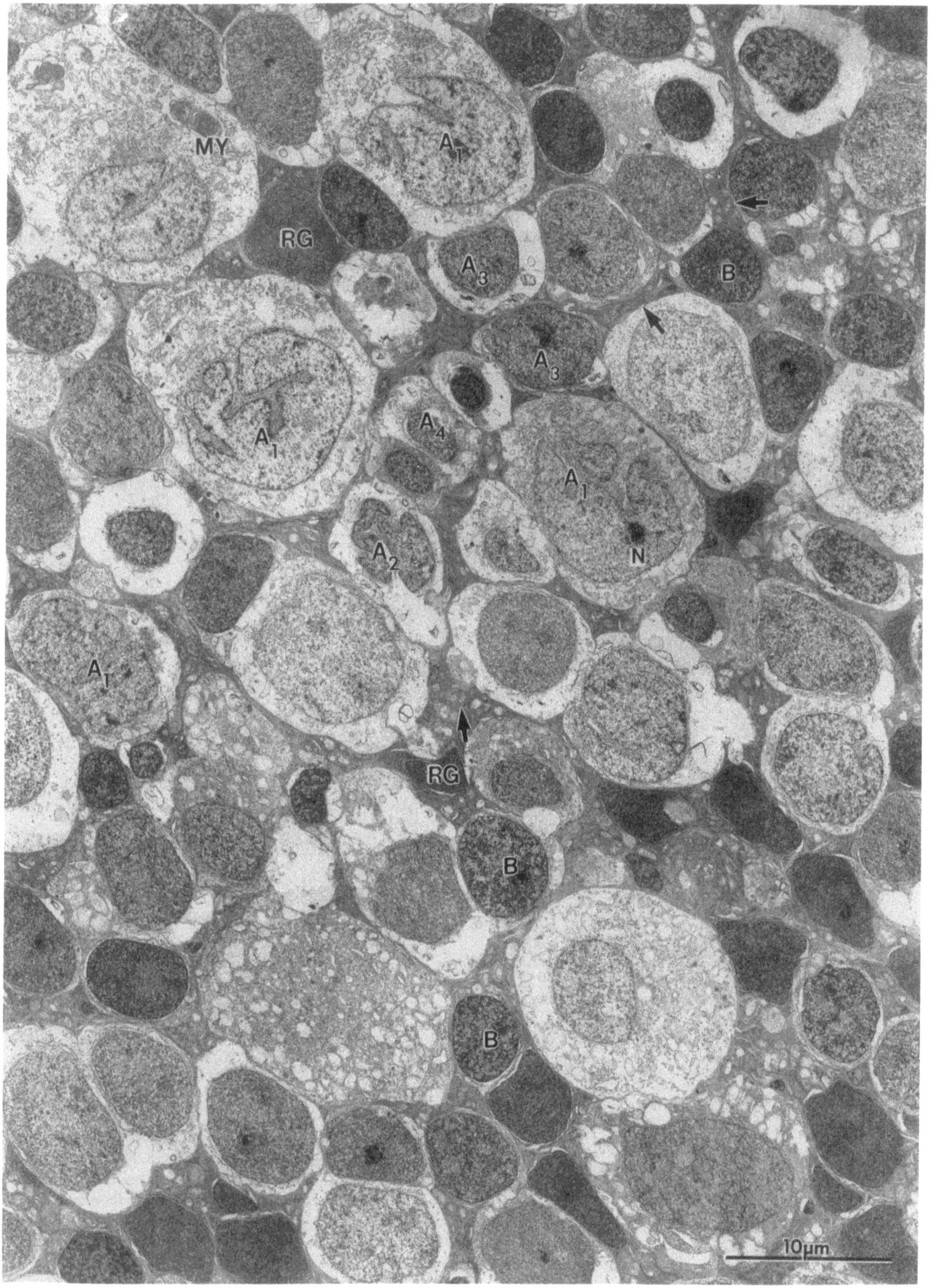
MY
A_1
RG
A_3
B
A_3
A_4
A_1
A_1
N
A_2
A_1
RG
B
B
10μm

The perikaryon of a small amacrine cell is shown in **Figure 39**. It is taken from a tangential section through the peripheral retina of a cynomolgus monkey. The level of amacrine cells in the internal nuclear layer contains a planar net of blood capillaries. A capillary is seen in longitudinal section in the lower part of Figure 39, that holds two erythrocytes (ER). Its wall is continuous. The endothelial cells (E) are joined by an elaborate junctional complex (inset). Occasionally pericytes (P) are seen in the wall of capillaries. The external leaflet of the vascular basilar membrane has two dense lines in cross section (arrowheads). The outer one represents the basilar membrane of the radial glial cell.

Radial glial cell extensions (RG) fill the space between the neuronal and vascular structures. The neurons are separated from the capillary wall by thin sheaths of glial cell cytoplasm (straight arrow). Some axonal profiles (A) of bipolar cells, 0.5 to 0.6 μm across, are seen. They are ensheathed by a mesaxon of the radial glial cell membrane (curved arrow).

The perikaryal profile of the amacrine cell in the center of Figure 39 is 6.5 to 8.3 μm in diameter. It has a kidney shaped nucleus with small irregular islets of electron-dense heterochromatin against a clear background. The cytoplasm is rich in organelles, such as rough endoplasmic reticulum (RER), free ribosomes (R), mitochondria (M), Golgi apparatus (G), and small lysosomal dense bodies (D).

Figure 39. Amacrine cell and blood capillary in tangential section through peripheral retina of cynomolgus monkey, ×16,800; inset, ×28,000. A, axon; BM, basilar membrane; D, dense bodies (lysosomes); E, endothelial cell; ER, erythrocyte; G, Golgi apparatus; M, mitochondrion; P, pericyte; R, free ribosomes; RER, rough endoplasmic reticulum; RG, retinal radial glial cell; straight arrow, cytoplasmic sheath of glial cell cytoplasm; curved arrow, mesaxon; arrowhead, dark sublamina of basilar membrane.

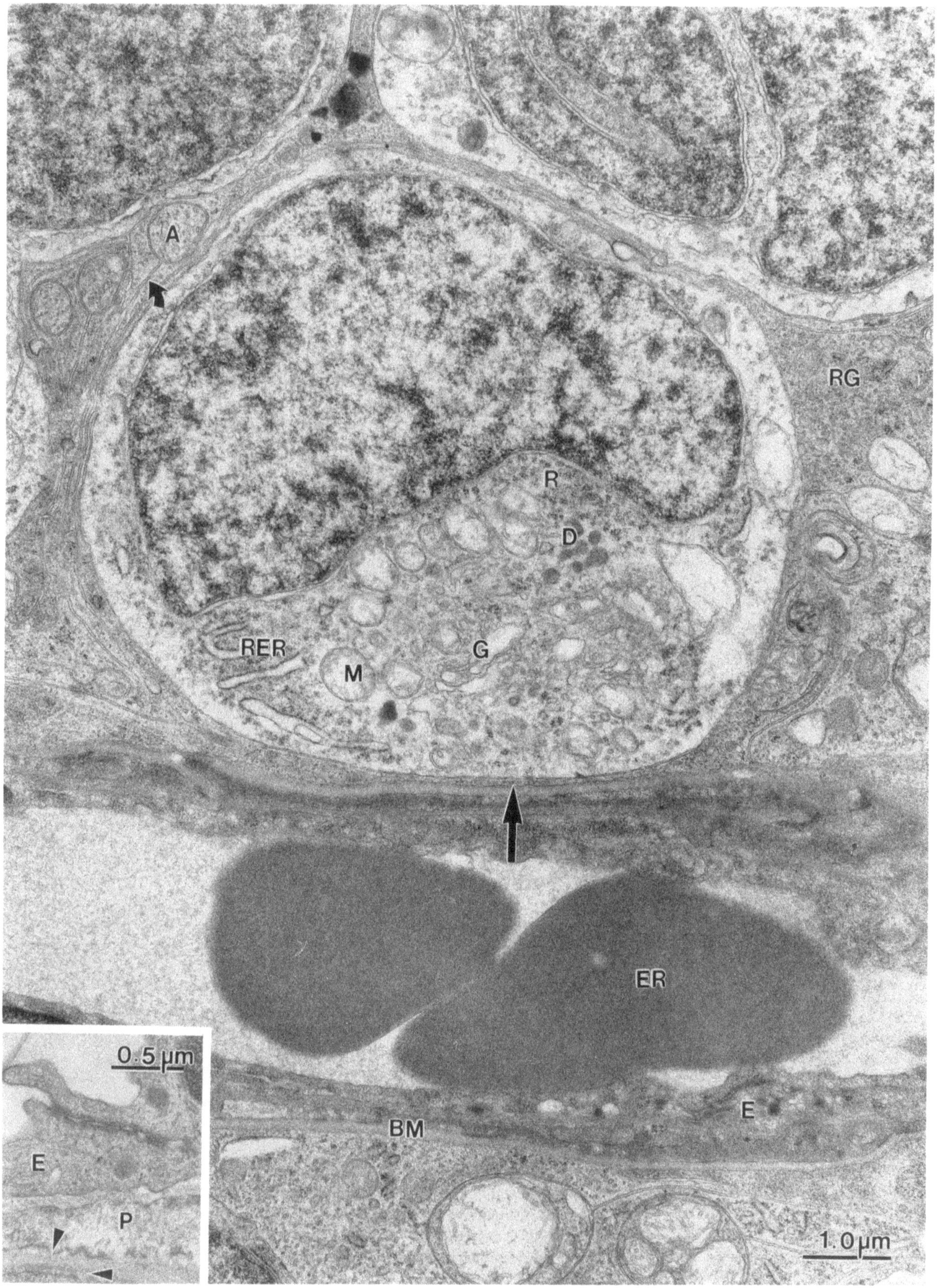
A
RG
R
D
RER
G
M
ER
0.5 µm
E
P
BM
E
1.0 µm

Figure 40 depicts further details of amacrine cell fine structure. In Figure 40A three amacrine cells are seen lying in close apposition to each other. No glial cell cytoplasm is separating them. One cell (AC_1) contains dense bodies (D) that may contain catecholamines or neuropeptides. The nuclei of the cells designated as AC_1 and AC_2 contain evenly dispersed finely granular chromatin. The third cell (AC_3) has an indented nucleus with distinct islets of heterochromatin. The cytoplasmic space of the amacrine cells is filled with cytoplasmic organelles. Golgi fields (G) are almost always seen in the perikaryal cytoplasm. The rough endoplasmic reticulum (RER) forms regular stacks, seen under the light microscope as Nissl substance. The cytoplasmic space abounds with mitochondria (M). A junction between an amacrine and a radial glial cell is seen (J).

The amacrine cell depicted in Figure 40B features a nucleus (N) similar to that in the cells AC_1 and AC_2 in Figure 40A. However, its cytoplasm has a ground substance that is more electron-dense than that of those cells. The dense bodies (D) in the amacrine cell seen in Figure 40B are stained lighter in their center than in their periphery. There are two Golgi fields (G) in the perikaryal cytoplasm of that cell.

The profiles of the radial glial cells (RG) contain numerous glycogen granules (GL). Axons of bipolar cells (A) with their mesaxons (curved arrows) are seen in Figures 40A and 40B.

Amacrine cells frequently generate cilia (Fig. 40C) from a well-developed basal body (BB). The striated rootlet fiber does not stand out very clearly. The basal part of the cilium often is embedded into a pouch of the plasmalemma (arrows); the pouch seen in Figure 40C is 1.4 μm deep. The external part of the cilium is squeezed between the cells of the internal nuclear layer without any apparent freedom to move. It is not known why certain cells within the retina express cilia more than others.

Figure 40. Tangential sections through amacrine cells, cynomolgus monkey. (A) Peripheral retina, ×10,300. (B) Peripheral retina, ×12,300. (C) Parafoveal retina, ×35,800. A, bipolar cell axon; AC_1, AC_2, AC_3, amacrine cells; BB, basal body; C, centriole; D, dense bodies; G, Golgi apparatus; GL, glycogen; J, junction; L, lysosome; M, mitochondrion; N, nucleus; RER, rough endoplasmic reticulum; RG, radial glial cell cytoplasm; arrows, orifice of ciliary pouch; curved arrow, mesaxon.

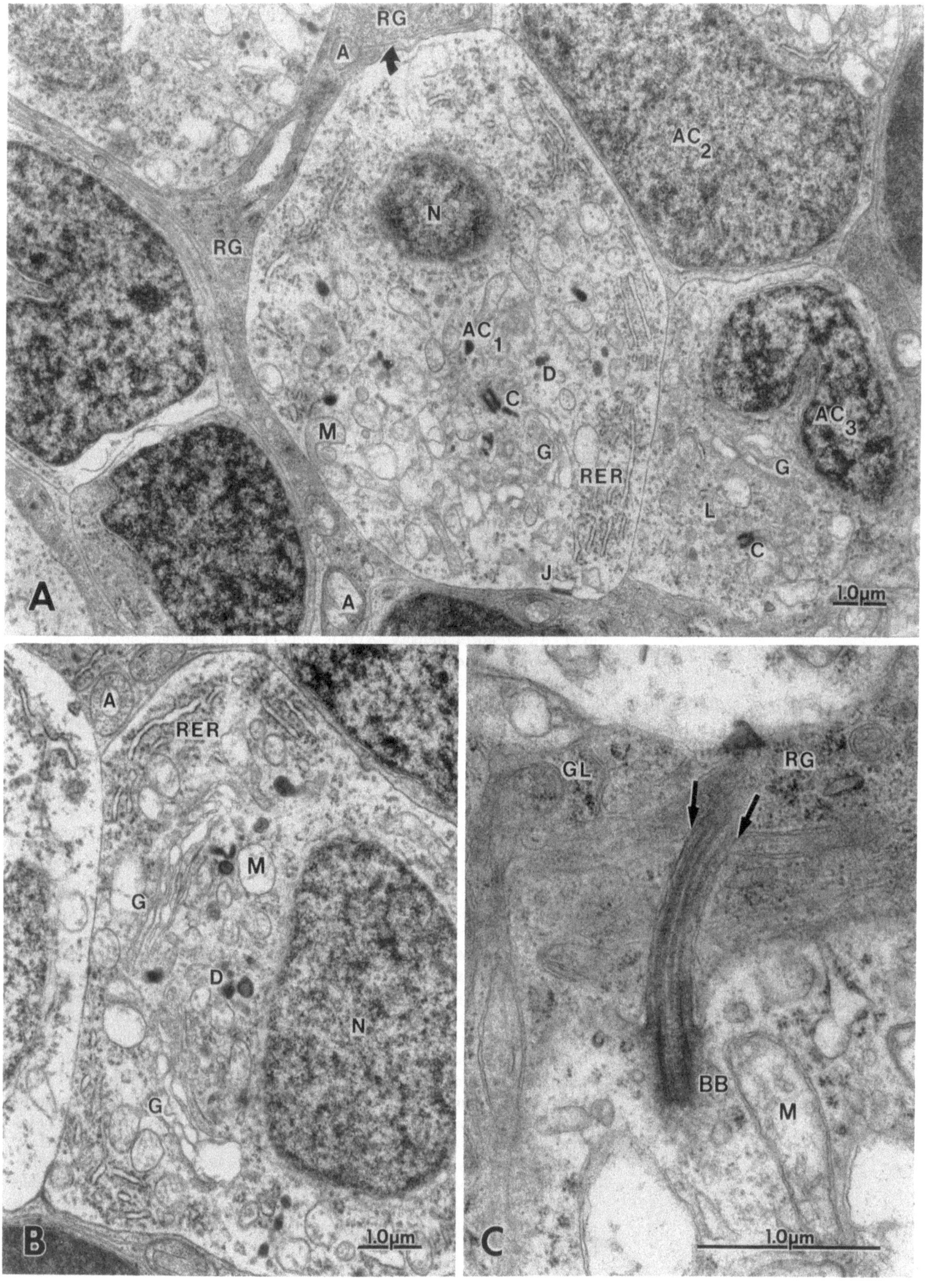
RG
A
RG
N
AC2
AC1
D
C
M
G
RER
AC3
G
L
C
J
A
1.0μm
A
A
RER
M
G
D
N
G
1.0μm
B
GL
RG
BB
M
1.0μm
C

Figure 41 shows some examples of junctional contacts between retinal cells. Figures 41A through 41D depict junctions made by amacrine cells within the inner nuclear layer. Gap junctions (GJ) as seen in Figure 41A were found in this material between amacrine cells within the inner nuclear layer. The profile of the gap junction shown is 0.8 μm long. There is some electron-dense material in the cytoplasm at both sides of the junction. The presence of gap junctions between amacrine cells implies that these cells are coupled electrically. It remains to be shown which of the many types of amacrine cells communicate with each other in this way in primates. In the inner plexiform layer of the cat retina, gap junctions have been seen between adjacent AII amacrine cell dendrites and between AII amacrine and cone bipolar cells (Famiglietti and Kolb 1975).

Amacrine cells form peculiar junctions associated with subsurface membranes (Figs. 41B–41D). These junctions can be to other amacrines (Fig. 41B), to bipolar (Fig. 41C) or to radial glial cells (Figs. 40A, 41C, and 41D). An electron-dense flattened membrane vesicle, the subsurface membrane (SM), lies adjacent and parallel to the junctional membrane in the cytoplasm of the amacrine cell. In Figure 41C this membrane is seen as continuous with the endoplasmic reticulum. In some instances several flattened membrane vesicles form a regular stack (Fig. 41D). At the curved arrow in Figure 41D, the uppermost membrane of such a stack is studded with some ribosomes. The significance of these structures is not known. Similar structures have been reported in ganglion cells of humans, monkeys, and rabbits (Nomura 1976).

Figures 41E through 41H show some synaptic contacts between neurons in the inner plexiform layer. A dyad synapse is typical of bipolar cells (Dowling 1987, pp. 48 ff). The synaptic ribbon (SR) depicted in the dyad synapse of Figure 41E, is 0.3 μm wide. It is surrounded by synaptic vesicles (SV). An arciform density (AD) is between the ribbon and the synaptic membrane at the synaptic ridge. There are regularly two postsynaptic endings (PS), the dyad, associated with the ribbon synapse in bipolar cells.

Figure 41. Junctional contacts within inner part of internal nuclear layer and within internal plexiform layer. (A) Gap junction, peripheral retina, cynomolgus monkey, ×74,000; (B–D) Junctions associated with subsurface membranes, peripheral retina, cynomolgus monkey. (B) ×74,000. (C) ×60,000. (D) ×46,000. (E) Dyad, peripheral retina, cynomolgus monkey, ×96,000. (F) Reciprocal synapse, parafoveal retina, baboon, ×60,000. (G) Dyades and conventional synapse in same terminal, peripheral retina, baboon, ×24,000. (H) conventional synapse, peripheral retina, cynomolgus monkey, ×40,000. AC, amacrine cell; AD, arciform density; BP, bipolar cell; DV, dense core vesicle; GJ, gap junction; M, mitochondrion; PS, postsynaptic ending; RG, radial glial cell; SM, subsurface membrane; SR, synaptic ribbon; T, microtubules in bipolar cell; V, synaptic vesicles; thin arrow, continuity between subsurface membrane and endoplasmic reticulum; curved arrow, ribosomes attached to subsurface membrane; thick arrows, direction of synapse; arrowhead, conventional synapse, bipolar cell being presynaptic.

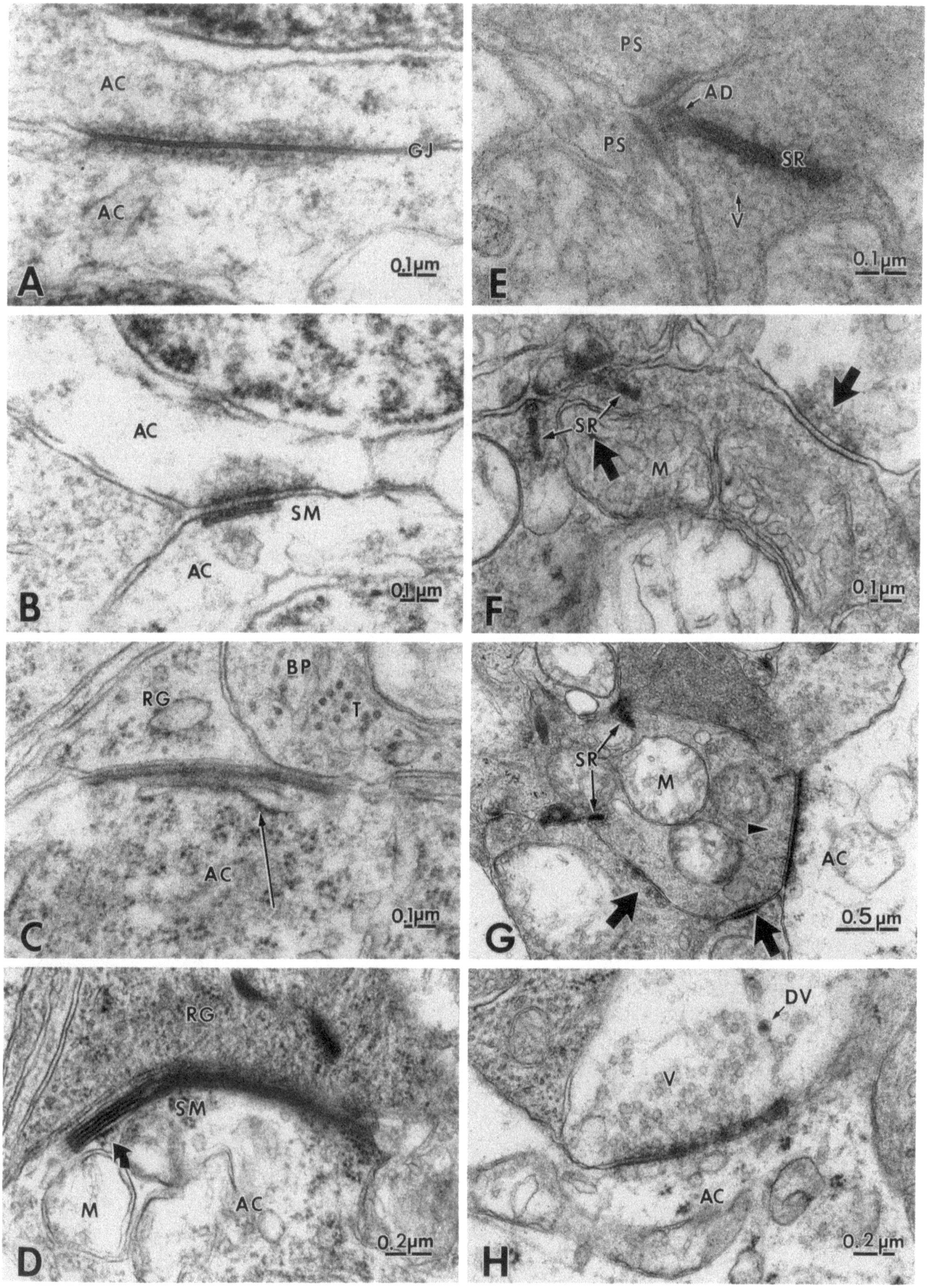
AC
GJ
AC
0.1µm
A
AC
SM
AC
0.1µm
B
BP
RG
T
AC
0.1µm
C
RG
SM
M
AC
0.2µm
D
PS
AD
PS
SR
V
0.1µm
E
SR
M
0.1µm
F
SR
M
AC
0.5µm
G
DV
V
AC
0.2µm
H

These endings may belong to a ganglion and an amacrine cell or to two amacrine cells (Dowling 1987, p. 66).

Figure 41F shows a section of a bipolar cell axon to which an amacrine cell is presynaptic (right thick arrow). In the axon, about 1 μm from this synapse, two synaptic ribbons are the presynaptic elements of two small dyads. This arrangement is called reciprocal synapse. It is believed to facilitate local feedback interactions between amacrine and bipolar cells (Dowling 1987, p. 67).

Another synaptic arrangement is seen in Figure 41G. A bipolar cell axon with two small synaptic ribbons (SR) in one location is the presynaptic partner of a conventional synapse (arrowhead) in another part of its membrane. The postsynaptic ending of the conventional synapse is an amacrine cell. Thus, bipolar cell axons may form both ribbon and conventional synapses. Two small synapses (thick arrows) are arranged reciprocally to those described earlier.

The last example (Fig. 41H) depicts a conventional synapse to an amacrine cell (AC). The presynaptic ending contains clear synaptic vesicles and one dense core vesicle (DV). This ending may belong to a peptidergic amacrine cell (Höfelt et al. 1980).

Ganglion Cells and Internal Limiting Membrane

Retinal ganglion cells receive signals from bipolar and amacrine cells. The synaptic contacts to these cells are made in the internal plexiform layer. Ganglion cells are as diverse in morphology as are amacrine cells. The morphology of their dendritic ramification and the size of their cell bodies are used to classify ganglion cells morphologically. However, the dendritic ramification pattern is not apparent in standard electron micrographs, and the ganglion cell type cannot be deduced from the following electron micrographs. Several classification schemes are discussed and correlated to each other in Rodieck (1973, pp. 469 ff).

Ganglion cell perikarya form a distinct stratum of the retina. It is comprised of a monolayer of perikarya in the peripheral retina. In its central part, ganglion cell bodies are stacked on top of each other, so that several rows of them are seen in radial sections of the retina. At the rim of the fovea, the ganglion cell layer is up to eight cell bodies deep. Axons of retinal ganglion cells make up the nerve fiber layer of the retina. They travel through the optic nerve to the visual centers of the brain.

Dowling (1987, pp. 77 ff) suggests the following classification of retinal ganglion cells regarding their connectivity: The cells of the first class receive input primarily from bipolar cells; the second class of ganglion cells is excited by bipolar and amacrine cells; and the third group is connected exclusively to amacrine cells.

Two classes of monkey ganglion cells, known as P (parvo) and M (magno) cells, are defined by their targets in the lateral geniculate nucleus of the thalamus (Shapley and Perry 1986). P cells have small receptive fields, contribute to color vision, and they have no contrast sensitivity. They connect to the parvocellular layers of the lateral geniculate nucleus. M cells do not contribute to color vision and they are sensitive to contrast. Their receptive field is larger than that of the P cells. They connect to the magnocellular layers of the lateral geniculate nucleus.

Within the retina, the ganglion cells are the terminal members of a variety of parallel neural pathways. Most of them are still poorly established. However, the principal rod specific pathway is worked out in detail. It transmits signals from rod cells via rod bipolar and the above mentioned AII amacrine cells to two types of ganglion cells, the on-center and the off-center ganglion cells (Daw et al. 1990). Another established pathway in the retina is described in Kolb and Lipetz (1991). This is the color-coding pathway for signals that originate in cones responding to light of long (red) or of intermediate (green) wavelength. The cone cells connect to flat, off-center bipolar cells, and to invaginating, on-center, midget bipolar cells. These cells make contacts directly to on- and off-center midget ganglion cells. Horizontal and certain amacrine cells modify the dynamics of these pathways.

Figure 42 is a light micrograph of a tangential semithin section through the peripheral retina of a cynomolgus monkey. At this retinal location the perikarya of ganglion cells form a monolayer. A planar network of capillaries (C) supplies blood to the ganglion cells.

The perikarya of ganglion cells differ in size. The largest profiles of them (GC_1) are between 22 and 26 μm in diameter. The cell bodies designated as GC_2 measure 14 to 16 μm, and those at GC_3 are between 8 and 9 μm wide. An even smaller perikaryon, approximately 5 μm in diameter, has a densely stained nucleus (arrow).

The distribution of the cell bodies appears to be at random. Adjacent ganglion cells often come into direct contact with each other. Along blood capillaries, the cell bodies of ganglion cells frequently are arranged in rows. The cytoplasm of the neurons is stained lightly. Their nuclei are round, and they contain a prominent dark nucleolus. A few nerve fibers (NF) are seen. Retinal radial glia (RG) fills the spaces between the neurons with densely stained cytoplasm.

Figure 42. Light micrograph of tangential semithin section through ganglion cell layer, cynomolgus monkey, peripheral retina, ×820. C, blood capillary; GC_1, GC_2, GC_3, ganglion cell perikarya; NF, nerve fibers; RG, cytoplasm of radial glial cell; arrows, small cell with dark nucleus.

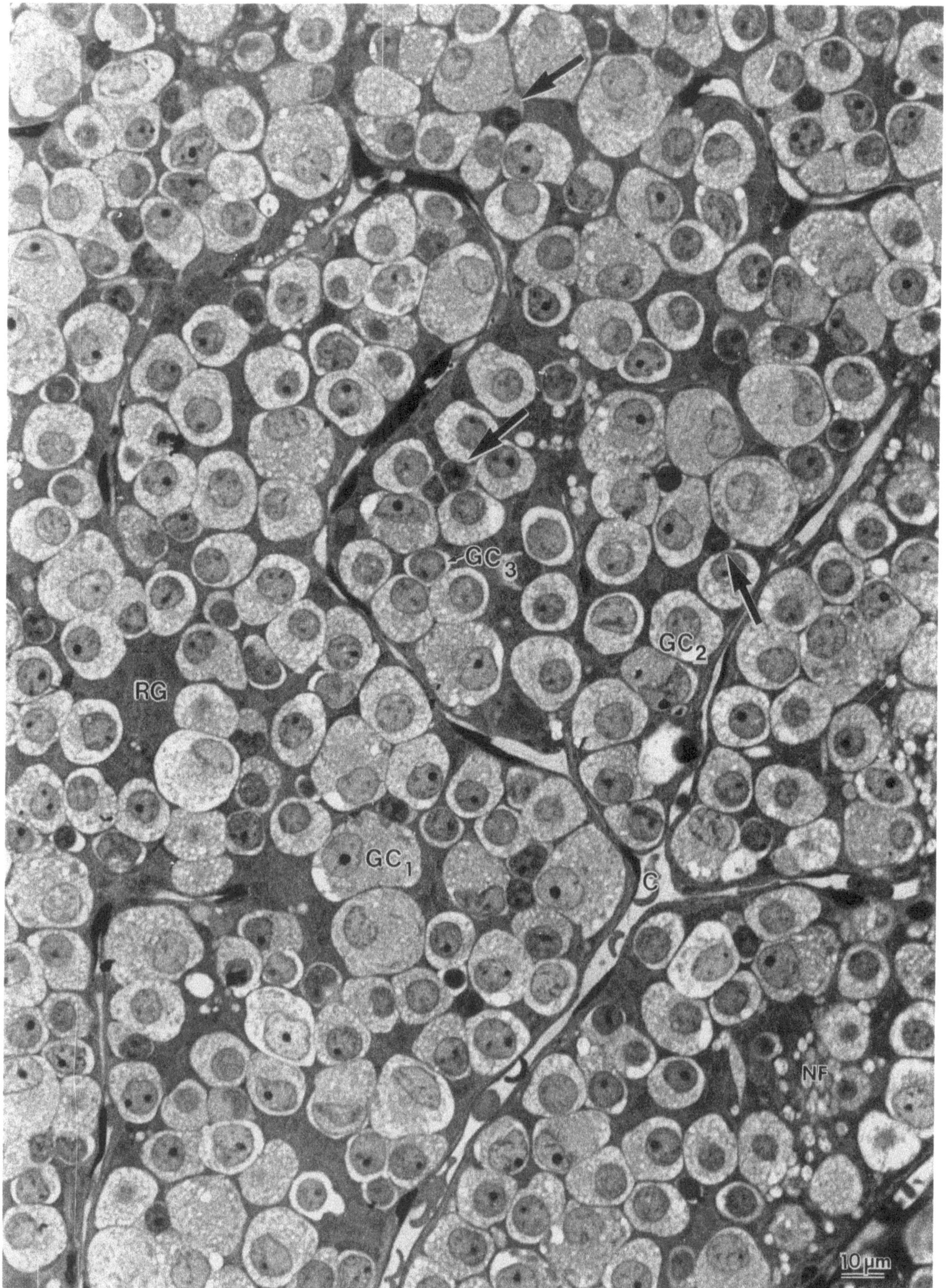

GC3
GC2
RG
GC1
C
NF
10 μm

Figure 43 is an electron micrograph of a tangential section at the same retinal location as shown in the previous light micrograph. The large ganglion cell perikarya (GC_1 and GC_2) contain nuclei with finely dispersed chromatin and a distinct, dark nucleolus (N). The cytoplasm is filled densely with organelles among which Golgi apparatus (G), mitochondria (M), and rough endoplasmic reticulum (RER) are most prominent. The small ganglion cell perikarya (GC_3) have a somewhat coarser chromatin in their karyoplasm. The very small ganglion cell designated as GC_4 corresponds to the cells seen at the arrows in Figure 42.

Some of the perikarya contain nuclei with deep indentations (asterisks). These cells may constitute displaced amacrine cells. However, an unquestionable classification of them cannot be made.

Radial glial cytoplasm (RG) is divided into irregular fields representing cross sections through the endfeet of retinal radial glial cells. Thin cytoplasmic sheaths of glial cytoplasm (arrows) separate the neurons from the walls of the blood vessels.

As elsewhere in the retina, the capillaries (C) have continuous walls. Occasionally, a pericyte (PC) is seen. The distance between capillary loops is approximately 30 μm.

Figure 43. Electron micrograph of tangential section through ganglion cell layer of peripheral retina of cynomolgus monkey, ×3,400. C, blood capillary; E, endothelium; EN, nucleus of endothelial cell; G, Golgi apparatus; GC_1, GC_2, GC_3, GC_4, perikarya of ganglion cells; M, mitochondria; N, nucleolus; PC, pericyte; RER, rough endoplasmic reticulum; RG, cytoplasm of radial glia; asterisk, perikaryon of presumptive displaced amacrine cell; arrows, cytoplasmic sheath of glia between neuron and blood vessel.

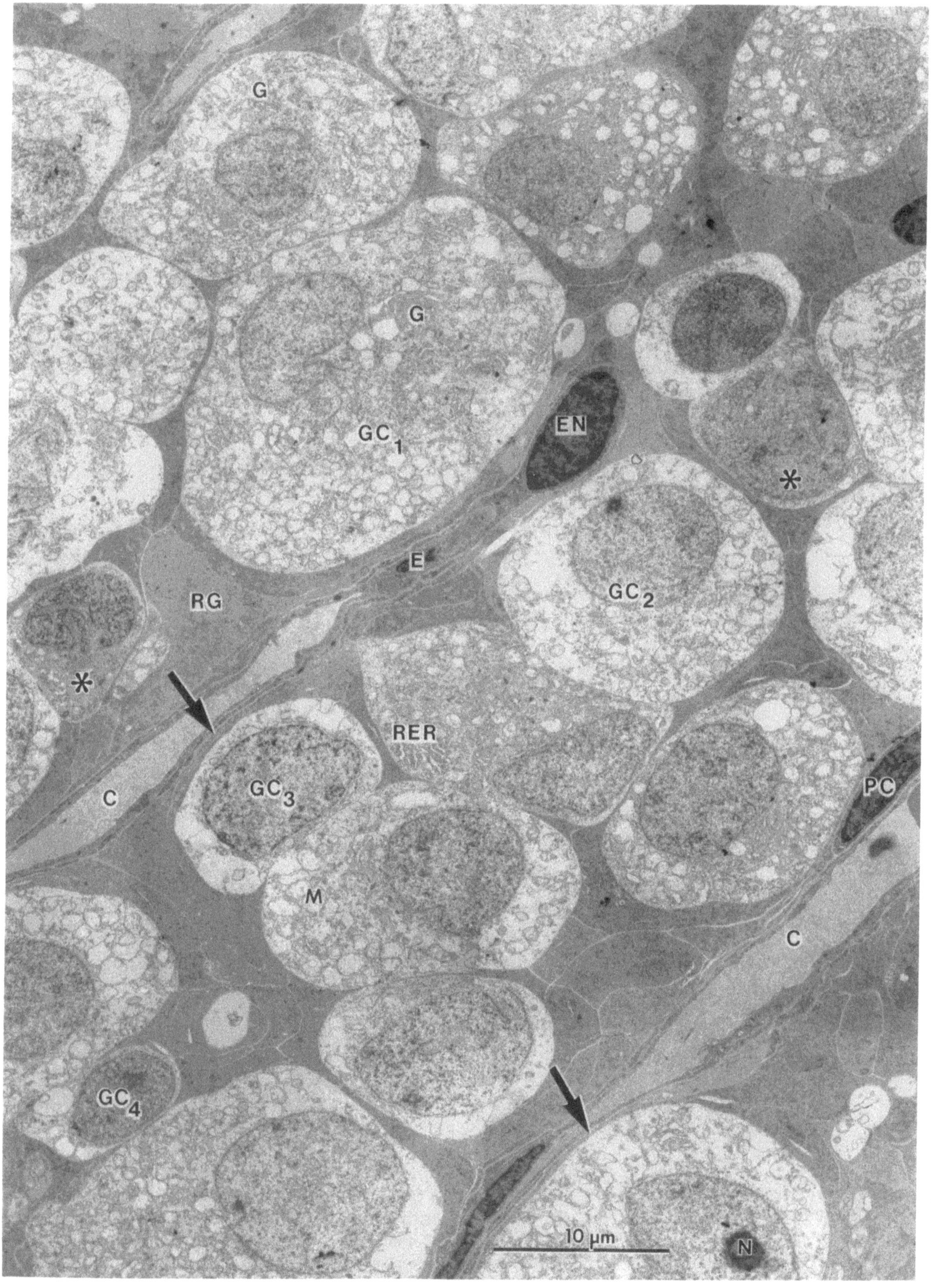
G
G
GC1
EN
E
RG
GC2
*
*
RER
GC3
C
PC
M
C
GC4
10 μm
N

Figure 44 shows a tangential section through the ganglion cell layer in the parafoveal region of the baboon retina. At this location the ganglion cell perikarya are so numerous that their nuclei are packed in several layers. Thus, not all profiles of neuronal cell bodies contain nuclei (asterisk). Mitochondria (M), in this example swollen because of slow fixation, are the major organelle in the ganglion cell cytoplasm. The large ganglion cell perikarya (GC_1) measure approximately 15 μm, the intermediate ones (GC_2) 10 μm and the small ones (GC_3, GC_4) 6 to 8 μm in diameter.

The nucleoplasm of the larger perikarya (GC_1, GC_2) contains finely granulated chromatin. The cell bodies designated as GC_3 contain a homogeneously and densely stained angular nucleus with a distinct nucleolus (N). The nucleus of the second kind of small perikarya (GC_4) is round and filled with islets of dark heterochromatin. The cytoplasm of the radial glia (RG) is confined to the narrow spaces between the ganglion cell perikarya.

Figure 44. Tangential section through ganglion cell layer of parafoveal baboon retina, ×4,000. GC_1, GC_2, GC_3, GC_4, perikarya of ganglion cells; M, mitochondrion; N, nucleolus; RG, radial glia; asterisk, cytoplasm of ganglion cell perikaryon nucleus of which is not in plane of section.

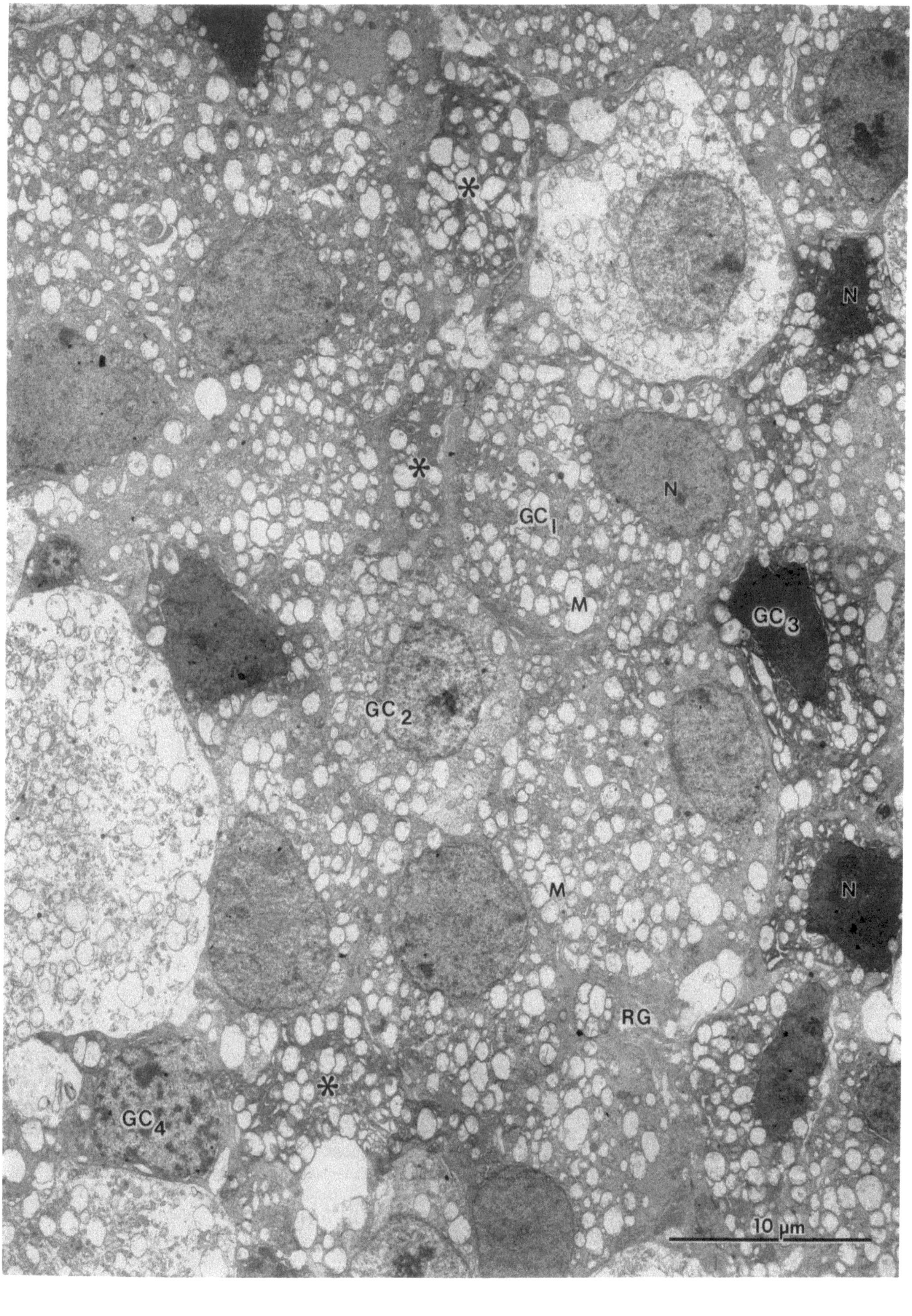

*
N
*
GC1
N
M
GC3
GC2
M
N
RG
*
GC4
10 μm

Figure 45 is from the interface between a ganglion cell and nerve fiber layer. The nerve fibers (NF) are the axons of the ganglion cells (GC). Neurotubules and mitochondria (M) dominate in their cytoplasm. The fibers often are in close contact to each other. Whenever they are further apart, cytoplasmic extensions of retinal radial glial cells fill the spaces between them.

The ganglion cell in the lower right of Figure 45A has a round nucleus with finely dispersed chromatin and a distinct nucleolus. Its cytoplasm is tightly filled with organelles. Mitochondria (M) abound. Blood capillaries (C) occur at the retinal level depicted here. A pericyte surrounded by basilar membrane is seen at P.

Endothelial cells (EN) of the cynomolgus monkey contain occasionally crystalloids as seen in Figure 45B. The crystalloid contains parallel rows of ring-like subunits, 30nm in diameter, and spaced 45 nm center-to-center. These inclusions have been reported in endothelia of ocular blood vessels in *Macaca mulatta* (Freddo et al. 1980). The chemical nature of the crystalloids and their function are not known.

Figure 45C is a section of the wall of a retinal artery. An endothelial cell (EN) is seen in the upper part of the picture. It contains dark round profiles 300 nm in diameter that might constitute cross sections through rod-shaped bodies known as Weibel-Pallade bodies (Freddo et al. 1980). By hemidesmosomes (arrowheads), the smooth muscle cells (SM) in the arterial wall are attached to the basilar membrane surrounding them.

Figure 45. Interface between ganglion cell and nerve fiber layer. (A) Cynomolgus monkey, peripheral retina, ×4,400. (B) Capillary wall, cynomolgus monkey, peripheral retina, ×68,000. (C) Arterial wall, baboon, parafoveal retina, ×10,000. BM, basilar membrane; C, blood capillary; CR, crystalloid; E, erythrocyte; EN, endothelium; GC, ganglion cell perikaryon; M, mitochondrion; N, nucleolus; NF, nerve fiber; P, pericyte; RG, radial glia; SM, smooth muscle cell; arrowhead, hemidesmosome; arrow, rod-shaped (Weibel-Pallade) body in cross section.

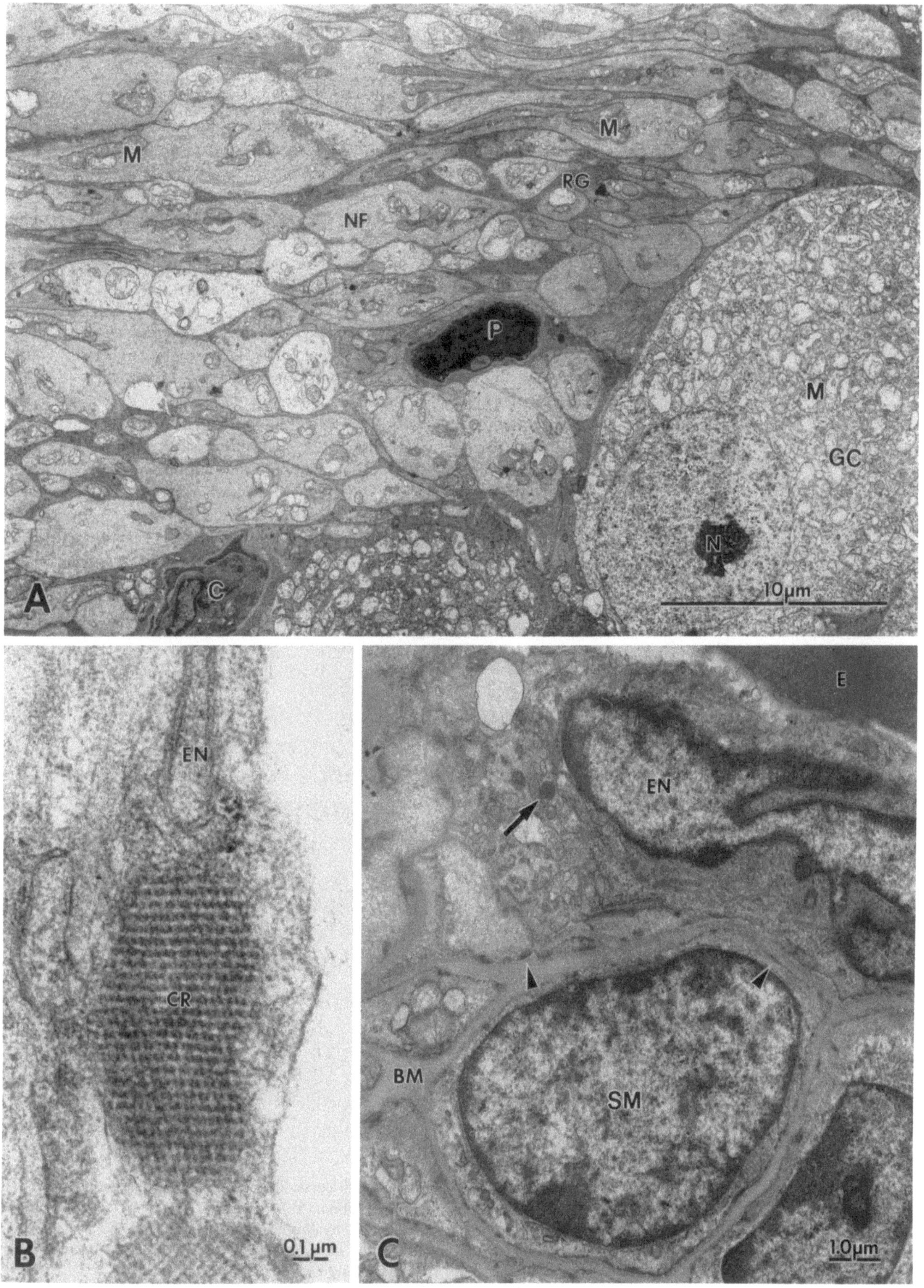
M
M
RG
NF
P
M
GC
N
C
10µm
A
EN
CR
0.1 µm
B
E
EN
BM
SM
1.0µm
C

Figure 46 shows radial sections through the nerve fiber layer of the baboon retina. In Figure 46A, the cross section of a nerve fiber bundle is seen that contains nerve fibers (A) of diameters ranging from 0.2 µm (curved arrow) to 3 µm. There seems to exist no specific order in the arrangement of the fibers within the bundle. The radial glia fills the spaces between them, but, the fibers are not completely insulated from each other; they touch their neighbors frequently. The nerve fiber layer of cynomolgus monkeys has been studied morphometrically by Ogden (1983, 1984). Ogden found the range of the fiber diameters to be between 0.2 and 3 µm, the same as just mentioned, and he determined the relative amount of glia in the fiber bundles as 20% to 30%.

The fiber bundles of the nerve fiber layer are separated from each other and from the ganglion cell layer by 3 to 4 µm wide septae of the radial glia (RG).

Between the nerve fiber layer and the ganglion cell layer extends a capillary network (C). In the upper part of Figure 46A are cell bodies of nerve cells (GC). Between the fiber bundles and the vitreous is a 6 to 7 µm fringe of glial cell extensions. Large extracellular vacuoles (V) are seen in this area of the retina. The vitreal surface of the retina is covered by a thin basilar membrane (arrow). The glial fringe and the basilar membrane are the morphological substrate of the internal limiting membrane.

On occasion, small free cells (HC) of the vitreal cavity are seen adjacent to the endfeet of the radial glia (Fig. 46B). These cells are called hyalocytes. Figure 46C depicts a nucleus adjacent to the surface of the retina. Its karyoplasm resembles that of the radial glial cells of the inner nuclear layer. Thus, it might be a displaced nucleus of a radial glial cell. However, astrocytes occur in this region of the retina, and it can be argued that this nucleus belongs to one of these cells.

Figure 46. Radial sections through nerve fiber layer and through internal limiting membrane, baboon, peripheral retina. (A) ×5,200. (B, C) ×6,600. A, axon; C, blood capillary; GC, ganglion cell; HC, hyalocyte; M, mitochondrion; N, nucleus; RG, radial glia; V, vacuolar cavities of extracellular space; straight arrow, basilar membrane of radial glia; curved arrow, smallest nerve fibers.

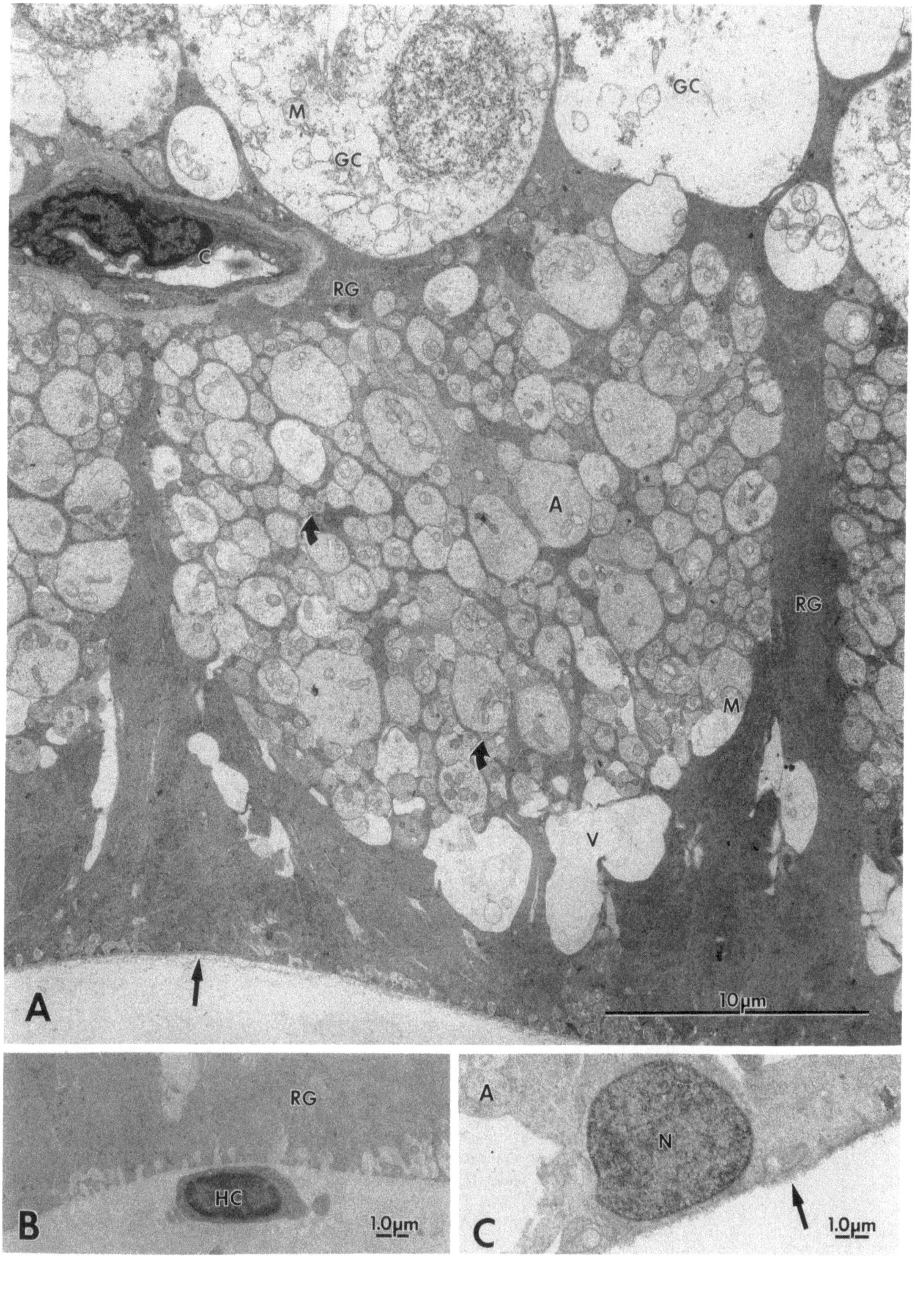
GC
M
GC
C
RG
A
RG
M
V
10μm
A
RG
HC
1.0μm
B
A
N
1.0μm
C

Figure 47 shows details of the inner limiting membrane in the parafoveal and foveal retina. Figure 47A is from a radial section, and Figure 47B is from a tangential section. In this region of the retina the basal cytoplasmic membrane of the radial glial cells is highly irregular. It almost forms a basal labyrinth similar to the basal plasmalemma of the retinal pigment epithelium. The membrane infoldings are embedded in an extracellular matrix containing basilar membrane (arrowhead) and fibrillar material (F). The surface of this matrix facing the vitreous is smooth (thick arrow). The basal membrane has focal densities (H) resembling hemidesmosomes.

The endfeet of the radial glial cells (RG) frequently are connected to each other by desmosome-like junctions (J). However, there is no junctional complex that could separate the extracellular compartment of the retina from the vitreous; many adjacent basal extensions of the glial cells are free of junctions (thin arrows). A few axons (A) of retinal ganglion cells are seen in Figure 47A; they contain mitochondria (M) and neurotubules. The cytoplasm of the glial endfeet is electron-dense, and it is poorly furnished with organelles.

Figure 47. Inner limiting membrane. (A) Radial section, parafoveal retina, baboon, ×14,300. (B) Tangential section, fovea, cynomolgus monkey, ×35,000. A, axon; F, fibers; H, hemidesmosome; J, desmosome-like junction; M, mitochondrion; RG, radial glia; thin arrows, endfeet of radial glial cells not connected to each other by junctions; thick arrow, surface of dense extracellular matrix; arrowhead, basilar membrane.

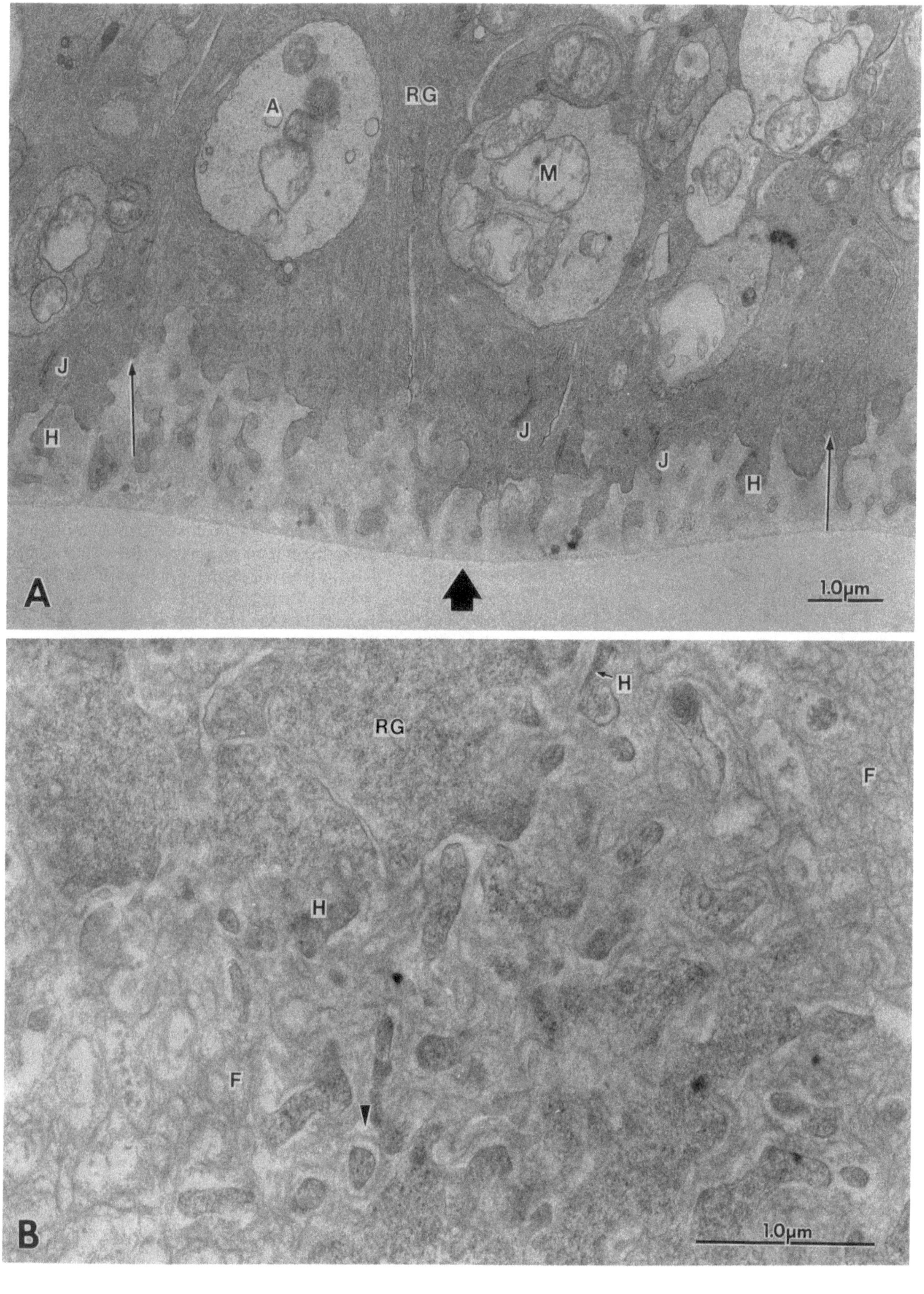
A
RG
M
J
H
1.0µm
B
F
1.0µm

Spatial Density of Retinal Cells

Information about the image received by photoreceptor cells flows through the retina via parallel channels. Each channel codes specific qualities of the visual information, such as color, brightness, or contrast. It is assumed that the first stages of these channels are at the level of bipolar cells (Gouras 1991). Models of retinal functions must consider the quantitative cellular composition of this tissue.

Table 1 summarizes the data obtained from four cynomolgus monkeys and the baboon. Some of these data have already been published (Krebs and Krebs 1987, 1989a). As described above (p. 12), the cell counts were done on electron microscopic montages of tangential sections at the levels indicated in Figure 6.

In the fovea, the pigment epithelium and cones have been counted in the foveolar center; the pedicles and the other retinal cells were counted 500 μm from the center (see Fig. 7). At this location, horizontal, bipolar, and amacrine cells are only slightly displaced from the pedicles to which they are connected. However, ganglion cells, being further displaced than the other neurons, receive input from bipolar and amacrine cells that are more centrally located. In addition, the ganglion cell layer at 500 μm eccentricity is as much as seven cell bodies deep. Not all the ganglion cells may be detected in the tangential sections, because their axons may course tangentially at a level different from that of the section. In the parafoveal sample, the ganglion cell layer still is several cell bodies deep and its assessment is equally problematic. Because of these ambiguities, foveal and parafoveal ganglion cells have not been counted. In the peripheral retina, the ganglion cells form a monolayer and pose no problem (Figs. 42 and 43).

Amacrine cell perikarya form a monolayer at the inner surface of the internal nuclear layer. They were counted in tangential sections of this level; their typical morphology allows identification (see Figs. 38–40). Bipolar cells were assessed in the central part of the inner nuclear layer. At this level either their axons, their dendrites, or their perikarya are in the plane of the section (see Figs. 33–35), and thus, no bipolar cell can escape detection. Some profiles in tangential sections of this level may belong to interplexiform cells. However, these cells are so rare that Dowling (1987, p. 29) found only a few more than 100 of them in several retinae, and the contamination of the bipolar cell counts by interplexiform cells is negligible.

Radial glial cell density was determined in tangential sections through their perikarya (see Fig. 36). Horizontal cells are easily identified. They form a monolayer of perikarya at the outer surface of the internal nuclear layer (see Fig. 31). As cone cell pedicles are restricted to a well defined layer at

Table 1. Spatial density of retinal cells.[a]

		Cells/mm² retinal surface area Cynomolgus monkeys[b] #1	#2	#3	#4	Baboon[b]
Pigment epithelium (P)	Peri.	3,500	3,330	4,600		3,000
	Para.			5,000	4,500	3,600
	Fovea	5,600	4,900	7,000		5,300
Cones (C)	Peri.	8,200	8,200	5,500		6,100
	Para.			12,400	13,600	15,300
	Fovea	157,000	150,000	230,000		113,000[c]
Pedicles (CP)	Peri.			5,400		6,000
	Para.			16,700	18,400	20,300
	Fovea	36,300	31,300	31,500		29,000
Rods (R)	Peri.	168,000	115,000	140,000		131,000
	Para.			145,000	157,000	114,000
Horizontal cells (H)	Peri.	6,500	7,000	5,300		5,000
	Para.			12,000	14,000	11,600
	Fovea	19,700	25,100	19,000		20,000
Bipolar cells (B)	Peri.	49,500	51,500	42,000		38,400
	Para.			85,900	90,000	97,100
	Fovea		78,800	79,000		89,000
Amacrine cells	Peri.	12,000	11,000	12,600		12,900
	Para.			19,300	21,000	18,400
	Fovea		14,000	11,000		10,200
Radial glial cells	Peri.	15,000	17,000	17,800		12,300
	Para.			17,800	23,000	17,500
	Fovea		17,000	16,600		20,000
Ganglion cells (G)	Peri.	6,750	5,350	3,800		3,700

[a] Abbreviations: Para., parafovea; Peri., periphery (at 30 degrees nasal eccentricity).
[b] The peripheral and foveal values of monkeys #1 and #2 and the foveal values of the baboon have been published in Krebs and Krebs (1987, 1989a).
[c] This value of 113,000/mm² is the density of foveolar cone inner segments at a distance of 100 μm from the foveolar center. In the center was a group of about 500 cones, the outer segments of which revealed a density of 385,000/mm².

the outer border of the external plexiform layer (see Figs. 25 and 30), their density could be determined in tangential sections through this level. Rod and cone densities have been measured in tangential sections through their inner segments (see Fig. 18); pigment epithelial cells, in sections through their nuclei or through their junctional complexes (see Figs. 9B–11).

Table 1 shows that the differences between corresponding samples of the five animals are small. Exceptions are the values for the foveal cones in monkey three and in the baboon (Krebs and Krebs 1989a). The 230,000 cone/mm^2 in the fovea of the third monkey probably was overestimated because there was a slight compression of the fovea. In spite of the relatively low value of the foveal cone density shown for the baboon in Table 1, the foveal center of this specimen contained about 500 central cones of very high spatial density (385,000/mm^2). This high density of outer segments of the central foveal cones, also seen by others (Ahnelt et al. 1987), may help to explain the psychophysical phenomenon of hyperacuity (Krebs and Krebs 1989a). Photoreceptor and ganglion cell densities in humans and monkeys have been measured repeatedly in retinal whole mounts (e.g., Curcio et al. 1987, Østerberg 1935, Stone 1981). Curcio et al. (1987) have demonstrated the capricious behavior of the fovea in retinal wholemounts, and the reported large variability of the receptor density in the fovea may have been artefactual (Williams 1988).

In the five animals used in Tables 1 and 2, the spatial density of the pigment epithelial cells increases in the fovea. However the number of photoreceptors per pigment epithelial cell remains relatively constant (**Table 2**). As shown earlier (see Figs. 18 and 26), the cross-sectional area of foveal cone inner segments is smaller than that of the cone cell pedicles. Thus, the pedicles of foveolar cone cells cover approximately five times larger an area than their inner segment profiles (Table 2). This difference in space requirement of the two domains of the central cone cells was used to explain the existence (Krebs and Krebs 1989a) and the geometry of the fovea (Schein 1988).

In the parafovea, cones are spaced loosely and the interstices between them are filled by rods (see Figs. 19 and 20). The displaced pedicles of cone cells in a given location of the parafovea are packed more densely than the cones above them (Table 2), because more fibers of Henle terminate than originate in any location of the extrafoveal macula. At the rim of the macula this space deficit is balanced, and the ratio of cone density versus pedicle density becomes 1.

Horizontal cells increase in density dramatically in the central retina. In the fovea there are approximately 20,000 horizontal cells per square millimeter of retinal area, or 0.5 to 0.8 horizontal cells per cone cell pedicle (Table 2). Similar values have been reported by Misotten (1974) in human retinas and by Wässle (1990) in monkey retinas.

Table 2. Numerical ratios between retinal neurons.

		Cynomolgus monkeys				
		#1	#2	#3	#4	Baboon
(C+R)/P	Peri.	50	37	32	40	45
	Para.			31	38	36
	Fovea	28	31	33		21
C/CP	Peri.			1	1	1
	Para.			0.74	0.74	0.75
	Fovea	4.33	4.82	7.3		3.9
H/CP	Fovea	0.54	0.8	0.6		0.7
B/CP	Fovea		2.5	2.5		3
B (rod)	Peri.		31,000	22,000		20,400
	Para.			44,000		36,200
B/G	Peri.	7.3	5.6	11	9.7	10.4

Abbreviations and symbols: Para., parafovea; Peri., periphery (at 30 degrees nasal eccentricity); (C+R)/P, photoreceptors per pigment epithelial cell; C/CP, densities of cones versus density of cone cell pedicles; H/CP, horizontal cells per cone cell pedicle; B/CP, bipolar cells per cone cell pedicle; B (rod), number of bipolar cells that might be rod specific per square millimeter of retinal surface area; B/G, bipolar cells per ganglion cell.

There are 2.5 to 3 bipolar cells per pedicle of a foveal cone cell. This number is consistent with hypothetical retinal circuitries discussed by Gouras (1991). Schein (1988) estimated only 2 bipolar cells per pedicle. The number of rod-specific bipolar cells at the peripheral and parafoveal locations can be estimated under the assumption that the connectivity of central and peripheral cone cells is alike. The number of peripheral cones multiplied by the bipolar cell versus pedicle ratio (B/CP) would result in the number of cone-specific bipolar cells. The measured total, minus the calculated cone-related bipolar cell density in a given retinal column, would give the spatial density of rod-specific bipolars. The results of such calculations are shown in Table 2 as *B(rod)*. It turns out that, in the parafoveal retina, the possible number of rod-specific bipolar cells is higher than in the periphery. One may conclude that either rod cells have a greater rate of convergence in the periphery, or that the assumption of equal connectivity of peripheral and central cone cells is wrong.

Six to 11 bipolar cells converge to 1 peripheral ganglion cell. This figure implies that, at least in the peripheral retina, a 1:1 ratio between cone-specific midget bipolar and ganglion cells cannot exist.

Fine Structure of the Choroid

Figure 48 shows radial sections of the baboon choroid. It is approximately 95 μm deep in the foveal region, but only 55 μm in the periphery (see also Fig. 6). The basal labyrinth of the pigment epithelium is at the bottom of Figure 48 beneath Bruch's membrane (B). Two small drusen are seen at D (Fig. 48A). The choroidocapillaris (C) is embedded in a fiber system consisting of Bruch's membrane, connecting fiber columns (FC), and an external collagen fiber layer (F). This recently discovered fiber system of the choroidocapillaris may be important for the patency of the capillary sinus (Krebs and Krebs 1990). The thickness of the choroidocapillaris decreases from 10 μm in the fovea to 7 μm in the periphery.

Melanocytes do not enter the layer of the choroidocapillaris, but they are abundant in the rest of the choroidal tissue. The distinct concentration of melanocytes in the external part of the *stroma vascularis*, reported by Hogan et al. (1972, p. 377) and Torczynski (1982), is seen in this material in the human choroid only. The choroidal melanocytes are large cells. Profiles of melanocytes exceeding 50 μm in length are seen frequently in the electron micrographs. In a wholemount of the human choroid, Torczynski (1982) showed a light micrograph of an approximately 100 μm long, star-like melanocyte. The cytoplasm of these cells is completely obscured by densely packed globular and ellipsoidal melanosomes 0.3 to 0.4 μm in diameter.

Veins (V) and arteries (A) tend to be more concentrated in the interior half of the *stroma vascularis*, at least in the central region where the choroid is thick. Nerves (arrows) in the choroid are remarkably abundant. In electron micrographs of radial sections through the baboon choroid in the foveal region (e.g., Fig. 48A), the average nearest neighbor distance between nerve fiber profiles is 20 μm.

Lymphatic clefts (L) are wider in the peripheral choroid (Fig. 48B). Occasionally a nucleus of a lymphatic endothelial cell (LE) is seen. These cells have not been recognized as lymphatic in the past (Torczynski 1982). However, recent ultrastructural evidence (Krebs and Krebs 1988b) proved that they are lymphendothelial cells.

Discrete collagen bundles (arrowheads) are seen in the stroma vascularis of the peripheral choroid (Fig. 48B). Their spatial arrangement remains to be studied. The suprachoroid (SC), a layer of several thin cytoplasmic sheaths, is seen in Figure 48B. A plasmacell (PC) lies between the melanocytes.

Figure 48. Choroid of baboon, radial sections, ×2,450. (A) Foveal sample. (B) Peripheral sample. A, artery; B, Bruch's membrane; C, choroidocapillaris; D, drusen; F, external fiber layer of fiber system of choroidocapillaris; FC, fiber column; L, lymphatic spaces; LE, lymphendothelial cell; P, choroidal pigment cell; PC, plasmacell; S, suprachoroid; V, vein; arrows, nerves; arrowheads, collagen fiber bundles.

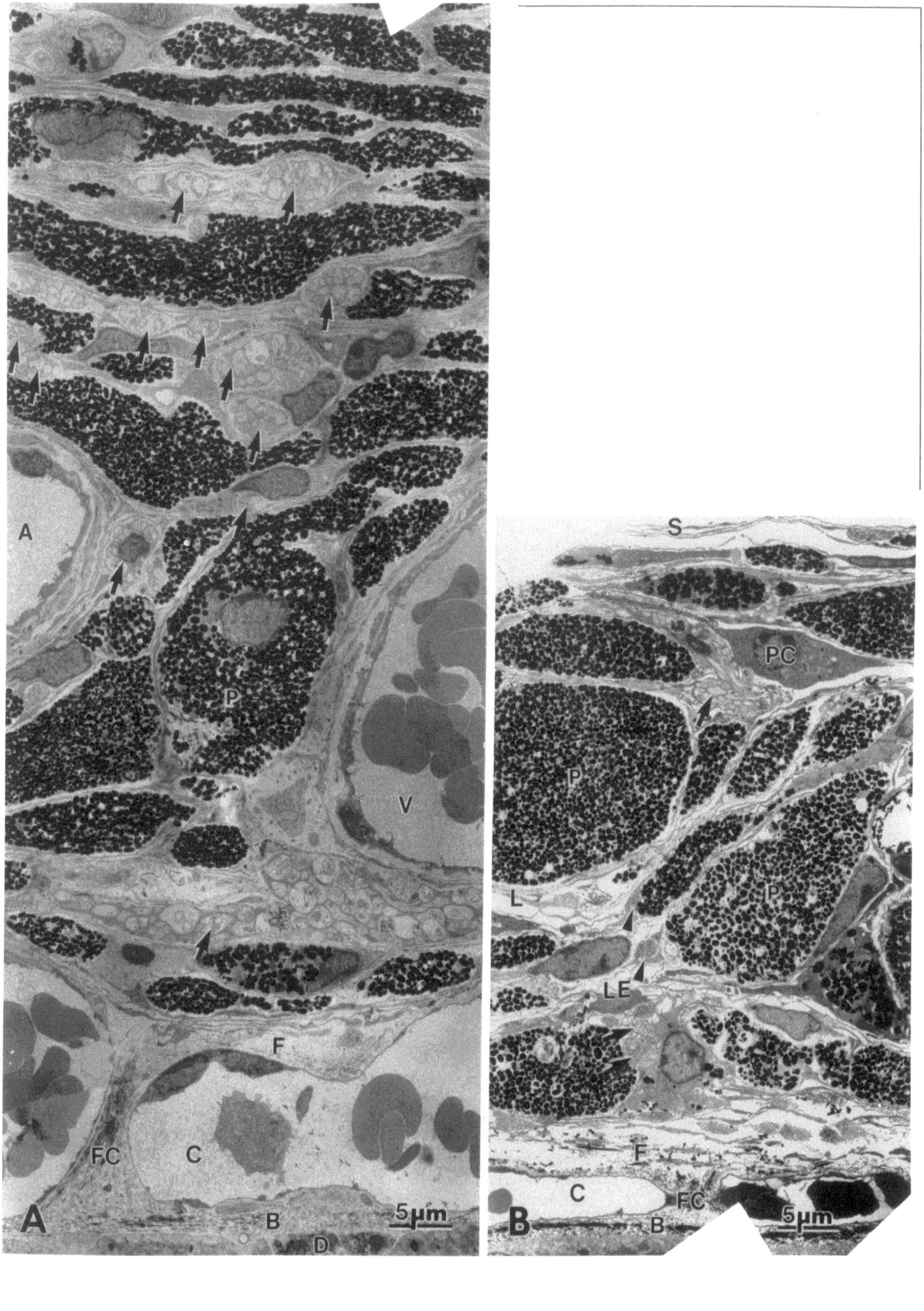
A
P
V
F
FC
C
B
D
5μm
A
S
PC
P
P
L
LE
F
C
FC
B
5μm
B

Choroidocapillaris and Its Fiber System

Bruch's membrane is regarded the innermost layer of the choroid. Its anatomical name, *complexus basalis*, refers to its complex structure. Commonly Bruch's membrane is subdivided in five lamina: (1) the basilar membrane of the retinal pigment epithelium, (2) an internal collagen layer, (3) a central elastic layer, (4) an external collagen layer, and (5) the basilar membrane of the choroidocapillaris. However, as the latter follows the circumference of the capillary sinus, its inclusion into Bruch's membrane does not seem logical (Krebs 1982). Bruch's membrane provides a smooth surface to which the retina is attached.

As was shown by Krebs and Krebs (1988b), Bruch's membrane is integrated into a fiber system that reinforces the choroidocapillaris. **Figure 49** is an artist's rendition of this system. Two flat leaflets of collagen and elastic fibers hold between them the capillary sinus of the choroidocapillaris. The internal leaflet is Bruch's membrane. The leaflets are connected to each other by straight fiber bundles passing through the tissue columns of the sinus. The fiber system is anchored in the vascular stroma of the choroid.

It has been suggested (Krebs and Krebs 1988b) that the fiber system prevents excessive expansion of the capillary sinus, because the distance between outer and inner leaflets is set by the length of the connecting fibers. In addition, the fiber system could act mechanically as a connecting link between the vascular stroma of the choroid and the retina.

Figure 49. Schematic presentation of fiber system of choroidocapillaris (not to scale). Stroma vascularis (SV), to which fiber system is anchored, is symbolized by one large artery (A). BL, basilar membrane of pigment epithelium; BM, Bruch's membrane; C, capillary sinus; CC, choroidocapillaris; CO, fiber columns; E, elastic layer of Bruch's membrane; F, external leaflet of fiber system. (From Krebs and Krebs 1990, by permission.)

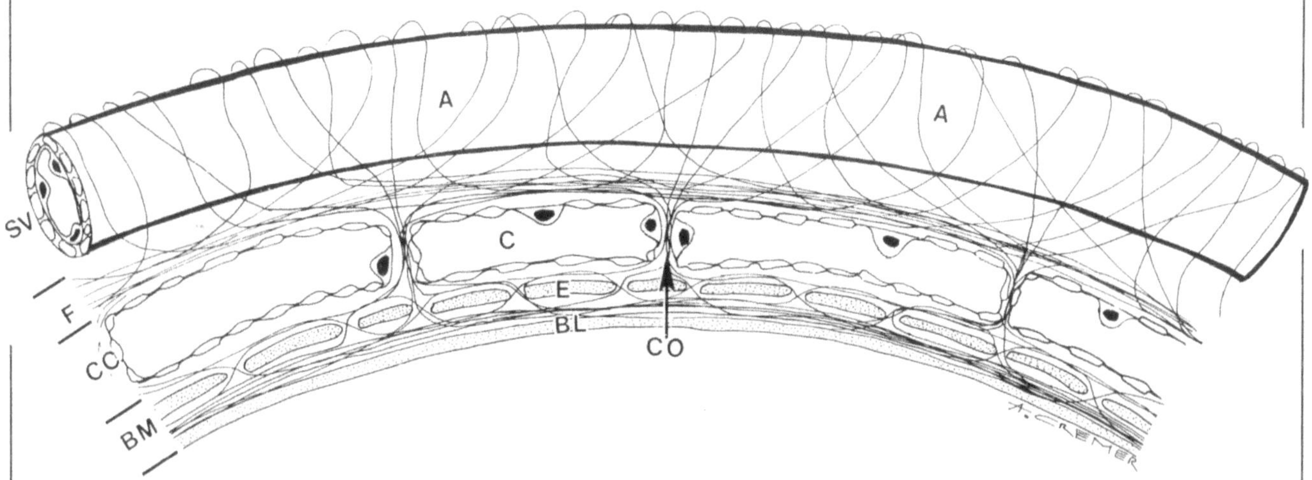
A
A
SV
C
F
E
BL
CO
CC
BM
A. CREMER

Figure 50 is a radial section through the choroidocapillaris in the foveal region of the baboon. The pigment epithelium is at the bottom. Its 200 nm thick basilar membrane (B), which does not follow the membrane infoldings of the basal labyrinth (BL), forms an even stratum. It is the first sublamina of Bruch's membrane (BM). Collagen fibers (C) and elastic material (E) make up most of Bruch's membrane (double-headed arrow).

Two profiles of the choroidocapillary sinus are seen in Figure 50, encircled by an external (basilar) membrane (BC). Their endothelial lining is studded with pores (thick arrows). A cell body of an endothelial cell (EC) contains an indented nucleus, densely stained cytoplasm with mitochondria, and rough endoplasmic reticulum. At the arrowheads are pinocytotic vesicles, indicating that the cell is actively taking up substances. These substances are transported in vesicles (straight arrow) to the luminal surface to be secreted into the lumen (curved arrows). Alternatively, this transport might be in the opposite direction. As the specimen has been fixed by immersion, erythrocytes fill the cavity of the sinus. A fiber column (CO) of the fiber system is cut longitudinally. The fibers of the column blend internally into Bruch's membrane and externally into the outer fiber leaflet (F).

Figure 50. Radial section through choroidocapillaris, foveal region, baboon, ×12,700. B, basilar membrane of retinal pigment epithelium; BC, basilar membrane of choroidocapillary sinus; BL, basal labyrinth of pigment epithelium; BM, Bruch's membrane; C, collagen fibers; CO, fiber column; E, elastic material; EC, endothelial cell; ER, erythrocyte; F, external leaflet of fiber system; M, mitochondrion; P, pigment granule; thin double-headed arrow, profile of Bruch's membrane; thick arrows, pores of capillary endothelium; straight arrow, transport vesicle; curved arrows, excretory vesicles; arrowheads, pinocytotic vesicles.

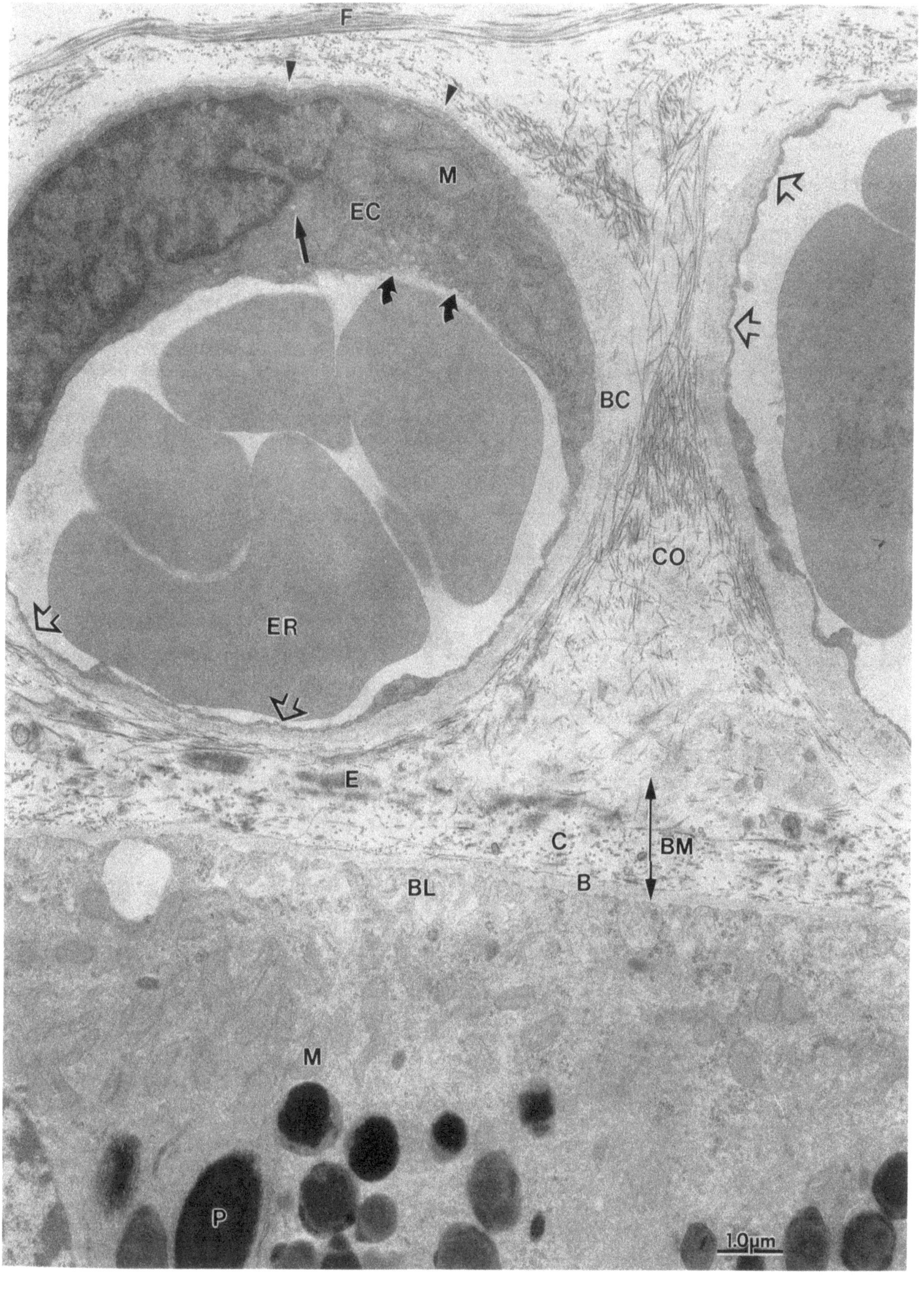
F
M
EC
BC
CO
ER
E
C
BM
B
BL
M
P
1.0μm

Figure 51 shows radial sections of Bruch's membrane in the peripheral human (Fig. 51A) and baboon (Fig. 51B) choroid and in the central choroid (Fig. 51C) of the baboon. Measured from capillary endothelium to pigment epithelium, Bruch's membrane is 1.5 μm to 1.9 μm deep in the peripheral human, 1.2 μm deep in the peripheral baboon, and 1.5 μm deep in the central baboon choroid. Thus, there might be a species difference. The structure of Bruch's membrane varies between the retinal loci shown in Figure 51. In the peripheral samples, the lamination of Bruch's membrane can be recognized. Adjacent to the basilar membrane of the retinal pigment epithelium is the inner collagen fiber lamina (CI). The central elastic lamina (E) is especially thick in the human sample (double-headed arrow). The external collagen lamina (CE) is thinner than the inner lamina.

In the central choroid of the baboon (Fig. 51C), the elastic material appears not to be confined to the central part of Bruch's membrane. It is found in small amounts in all layers, but in the external layers it is most abundant. This lack of a clear lamination of Bruch's membrane in the foveal and parafoveal regions was seen in all specimens described in this book.

Collagen fibrils from the connecting fiber columns are anchored in the internal collagen lamina of Bruch's membrane (arrows). Such anchoring fibrils have also been reported in the bovine choroid (Krebs 1982).

Figure 51. Radial sections through Bruch's membrane. (A) Human, peripheral choroid, ×21,300. (B) Baboon, periphery, ×26,500. (C) Baboon, parafoveal sample, ×13,400. BL, basal labyrinth of retinal pigment epithelium; BM, basilar membrane; C, lumen of choroidocapillary sinus; CE, external collagen lamina of Bruch's membrane; CI, internal collagen lamina; E, elastic material; EN, endothelium; N, nucleus; M, mitochondrion; arrows, collagen fibril of connecting fiber columns anchored in internal fiber lamina of Bruch's membrane; double-headed arrow, profile of elastic lamina of Bruch's membrane.

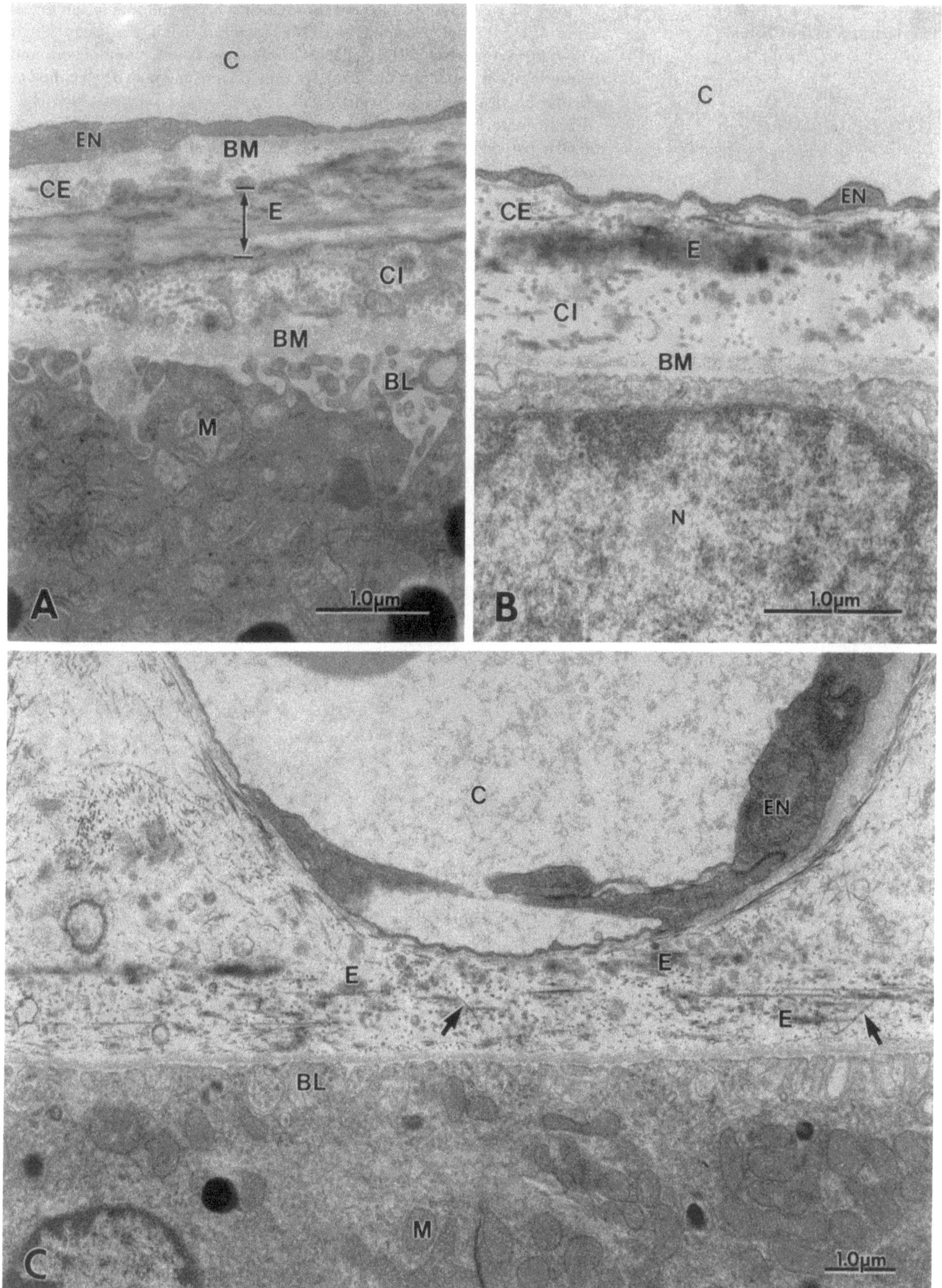
C
EN
BM
CE
E
CI
BM
BL
M
1.0μm
A
C
CE
EN
E
CI
BM
N
1.0μm
B
C
EN
E
E
E
BL
M
1.0μm
C

Figure 52 is a tangential section through Bruch's membrane in the foveal region of a cynomolgus monkey. Elastic material (E) and collagen fibrils (arrow) form a felt-like web. Collagen fibrils are straight, unbranched, and run randomly in any direction. They are thin (35 nm) and may be classified as reticular fibrils (Krebs and Krebs 1990).

Elastic material is seen as thick, branching strands, frequently arranged in star-like patterns. Between the fibers are lumps of amorphous, electron-dense material (M).

Figure 52. Tangential section through Bruch's membrane, cynomolgus monkey, foveal sample, ×6,600. E, elastic material; M, amorphous material; arrow, collagen fibril.

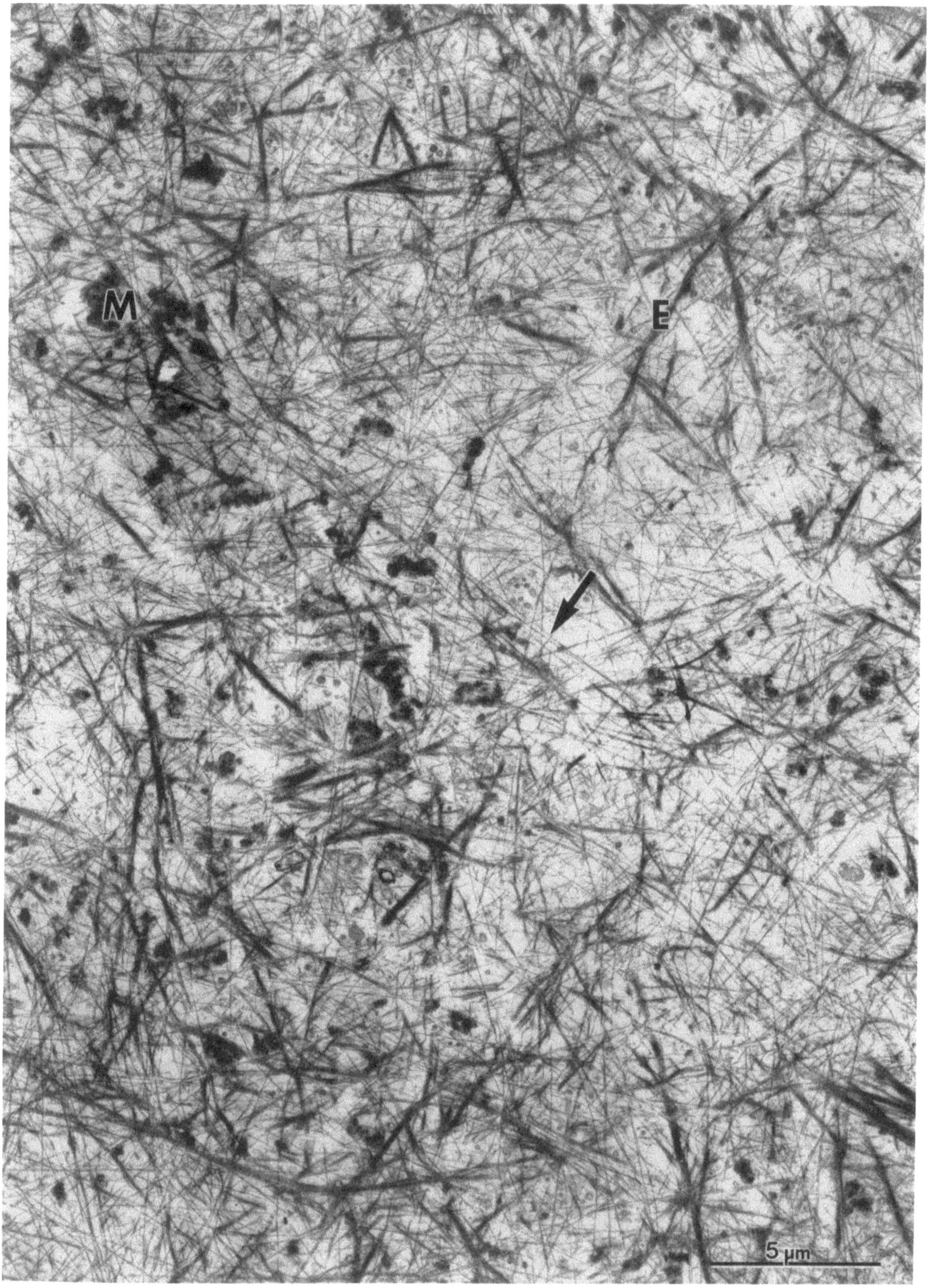
M
E
5 µm

Figure 53 depicts a tangential section through the choroidocapillary sinus in the foveal area of a cynomolgus monkey. The connective tissue columns (CO), connecting roof and bottom of the sinus, are cut transversely. They are covered with endothelium (E) and filled with fibrils. The smallest columns are approximately 1.5 μm and the larger ones 7 μm in diameter. A few erythrocytes (ER), two lymphocytes (L), and a granulocyte (G) are seen in the lumen of the sinus.

In the posterior choroid, which includes the peripheral samples of this material, the choroidocapillaris is sinusoidal. Anteriorly from the ocular equator, it becomes a dense network of capillaries (Hogan et al. 1972, p. 365). The arterioles filling the sinus and the draining venules are arranged in a way to result in a functional lobulation of the choroidal capillary sinus (Torczynski 1982).

As the choroidocapillaris supplies the retina with blood, it is not surprising that a number of intricate interactions occur between the capillary sinus and the retinal pigment epithelium. The pigment epithelium induces the pores in the capillary wall, and the choroidocapillaris provides essential clues for the polarization of the retinal pigment epithelium. For further information on this matter the review by Korte et al. (1989) may be consulted.

Blood flows through the choroid in such amounts that only a few percent of the arterial oxygen is extracted. It was suggested that this high blood flow helps to stabilize the temperature within the eye (Bill 1985). However, as Steinberg (1987) pointed out, the oxygen demand of the retinal pigment epithelium is so high that photoreceptors are forced to function at very low oxygen tension. As the latter also depend on oxygen supplied by the choroid, the high overall oxygen content of blood in the choroidocapillaris may be needed to provide a concentration gradient steep enough to drive sufficient amounts of oxygen into the retinal tissue.

Figure 53. Tangential section through choroidocapillaris in foveal region of cynomolgus monkey, ×2,500. CO, columns connecting roof and floor of capillary sinus; E, endothelium; ER, erythrocyte; G, granulocyte; L, lymphocyte; S, sinus lumen; arrow, small connecting column.

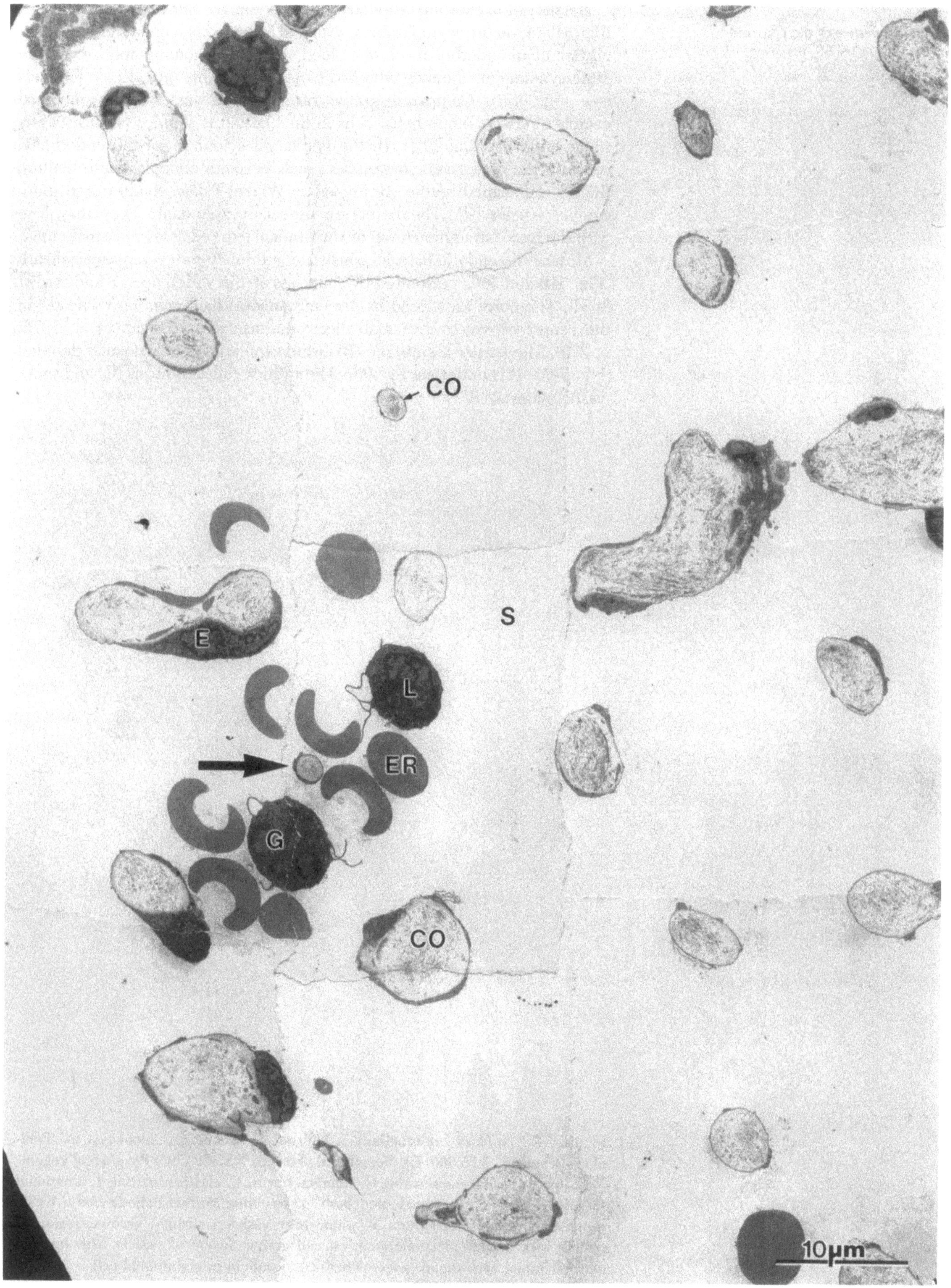
CO
S
E
L
ER
G
CO
10µm

Details of the choroidal capillary endothelium are shown in **Figure 54**. In Figure 54A, an intracytoplasmic crystalloid (asterisk) is seen in the perikaryal portion of an endothelial cell of a blood vessel. This micrograph is from the same cynomolgus monkey who had crystalloids in his retinal blood vessels (see Fig. 45B). Compared to the retinal blood vessels, the junctional complex between endothelial cells in the choroid is simple. It consists of a single adhering zonule (J). In the perikaryal cytoplasm of the endothelial cells are the cytoplasmic organelles, such as rough endoplasmic reticulum (RER), rod-shaped bodies (R) known as Weibel-Pallade bodies, and pinocytotic vesicles (V). The latter are especially abundant. They may fuse with the basal (straight arrows) or the luminal (curved arrow) plasmalemma.

Most of the endothelial wall consists of a thin (70 nm) cytoplasmic sheath (Fig. 54B and 54C) perforated by countless 50 nm wide pores (P and arrowhead). The pores are closed by a membraneous diaphragm (arrowhead), in the center of which is a small electron-dense mass (Hogan et al. 1972, p. 370). The basilar membrane (B) of the capillaries is occasionally doubled (Fig. 54A). It is underlain by 30 to 34 nm thick collagen fibrils (C) and some elastic material (E).

Figure 54. Capillary endothelium in choroid of cynomolgus monkeys. (A) Peripheral choroid, ×75,000. (B) Peripheral choroid, ×37,500. (C) Parafoveal region, ×41,300. B, basilar membrane; C, collagen fibrils; E, elastic material; J, junctional complex; P, pores; R, rod-shaped body, resembling Weibel-Pallade body; RER, rough endoplasmic reticulum; V, pinocytotic vesicles; straight arrows, fusion of vesicle with basilar plasmalemma; curved arrow, fusion of vesicle with luminal plasmalemma; arrowhead, pore; asterisk, crystalloid in endothelial cell.

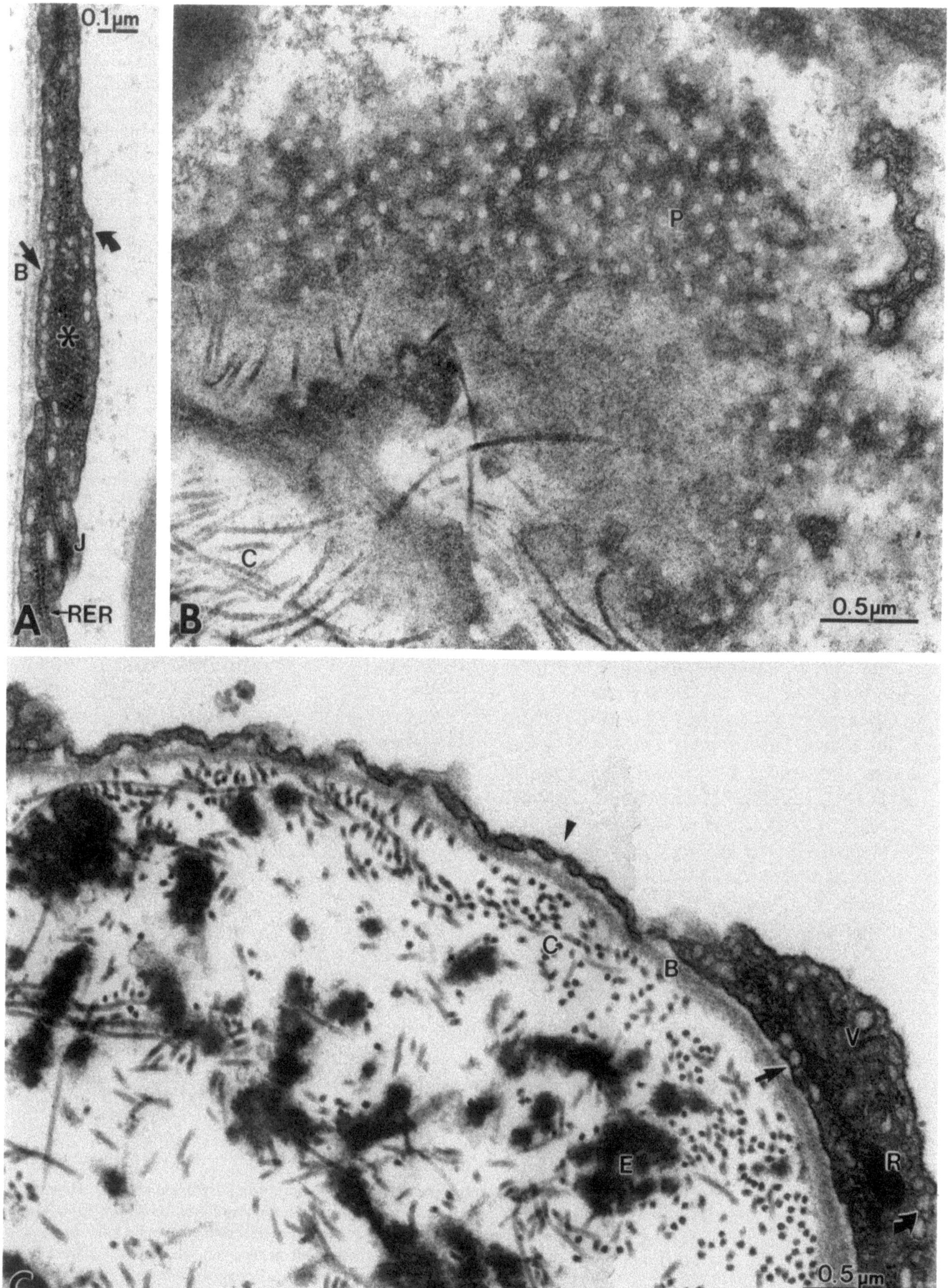
0.1μm
B
*
J
RER
A
P
C
0.5μm
B
C
B
V
E
R
0.5μm
C

Figure 55 depicts tangential sections through a connective tissue column of the choroidocapillaris and through a feeding arteriole in the parafoveal region of a cynomolgus monkey. The connective tissue column (Fig. 55A) is covered by the perforated endothelium of the choroidocapillaris. Pores are indicated by arrows. Two perikarya of endothelial cells (E) are seen. The basilar membrane of the endothelium is approximately 130 nm thick. Fibrocytes (F) are seen occasionally in the connective tissue columns. However, the most prominent content is the connecting collagen fiber bundles (CO) in the center of the column. Often the collagen fibers are accompanied by elastic material (EL).

In contrast to the large draining veins, the arterioles (Fig. 55B), feeding blood into the capillary sinus, have a relatively small diameter, 26 μm in this case. The endothelial perikarya (E), bulging into the lumen (L), are covered with microvilli. The endothelial wall still contains pores (arrows). Likewise, the veins near the choroidocapillaris have porous walls (Fig. 63A) in these specimens. Pores outside of, but close to, the choroidocapillaris support the hypothesis, reviewed by Korte et al. (1989), that a pore-inducing substance is secreted by the retinal pigment epithelium. If this substance diffuses beyond the outer surface of the choroidocapillaris it should induce pores in the endothelium of adjacent blood vessels.

Several slim pericytes (P) surround the endothelial tube of the arteriole. A fibrocyte (F) is seen in the neighborhood. The outer leaflet of the fiber system of the choroidocapillaris forms a thick sleeve of collagen fibers (C) around the blood vessel. Some elastic material (EL) lies between the collagen. Choroidal melanocytes (PI) are seen in the upper part of Figure 55B.

Figure 55. Tangential sections through choroid in parafoveal region, cynomolgus monkey. (A) Connective tissue column of choroidocapillaris, ×15,600. (B) Profile of feeding arteriole, ×3,600. B, basilar membrane; C, collagen fibrils; CO, connecting fibers of fiber system; E, endothelial cell perikaryon; EL, elastic material; F, fibrocyte; L, lumen; P, pericyte; PI, choroidal pigment cell; arrows, endothelial pore.

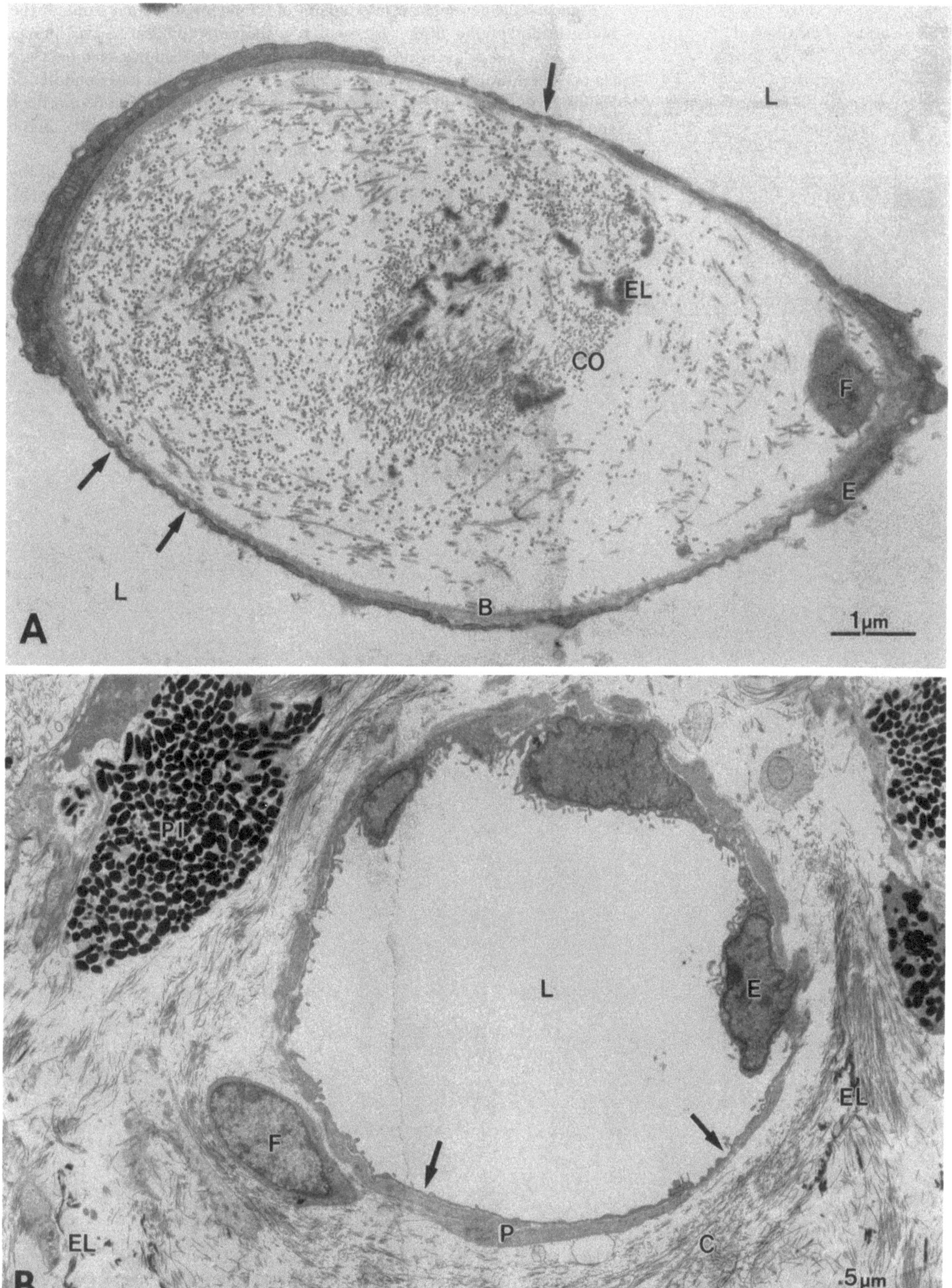

L
EL
CO
F
E
L
B
A
1µm
PI
L
E
EL
F
P
C
EL
B
.5µm

Figure 56 is an electron micrograph of a tangential section through the outer leaflet of the fiber system of the choroidocapillaris in the human specimen. In this stratum of the choroid, the sole cellular elements are sheath-like fibrocytes (F). Collagen fibers (C), arranged in loose bundles or singly, form a fabric-like sheath. The fibers are between 30 and 60 nm thick and display the typical cross striation of collagen (Fig. 56, inset). Some elastic material (E) is frequently seen among the collagen fibers.

The fiber system was only recently described as a structural unit of the choroid (Krebs and Krebs 1990). However, its components are clearly seen in many previously published micrographs (e.g., Hogan et al. 1972, Figs. 8-5, 8-7, and 8-28).

Figure 56. Tangential section through outer leaflet of fiber system of choroidocapillaris, human, 3,200; inset, ×102,500. C, collagen fibrils; EL, elastic material; F, cytoplasm of fibrocyte.

C
F
EL
10 µm
0.1 µm

Arteries, Veins, and Lymphatic Spaces

Figure 57 shows a section of the *stroma vascularis* of the cynomolgus choroid. An artery (A) with a thick *lamina elastica* (EL) and with an incomplete smooth muscle layer is surrounded by fibrous connective tissue. The periarterial tissue is free of melanocytes (P). It contains numerous small nerves (N) and sheath-like fibrocytes (F), subdividing the collagen fibers into several strata. The nerves seldom reach the arterial wall proper. They terminate without any specialization (arrows) between the collagen/fibrocyte lamellae close to the outer arterial muscle layer.

In the upper part of Figure 57 is a vein (V). Its wall is composed of endothelium and occasional pericytes (PC). There is no adventitia comparable to that of the artery. Melanocytes are seen frequently adjacent to the venous wall. Hogan and Feeney (1961) studied the choroidal blood vessels systematically and found that in the choroid, veins are larger and arteries are smaller in diameter than are comparable vessels in the general circulatory system.

Between artery and vein lies a large lymphatic space (L). Such spaces have been seen frequently by other workers, but they have been brushed off as artifacts. As the artery seen in Figure 57 has shrunk during fixation, the width of the lymphatic cleft may be wider than in vivo. However, the lumen of these spaces always is free of collagen and their lining consists of lymphendothelium (Krebs and Krebs 1988b). A lymphendothelial perikaryon (LE) is seen in the upper right in the wall of a lymphatic cleft. In its lumen is a macrophage (MP) filled with ingested melanosomes.

Figure 57. Radial section through peripheral choroid of cynomolgus monkey, ×3,200. A, artery; C, collagen; E, endothelium; EL, *lamina elastica*; F, fibrocyte; L, lymphatic space; LE, lymphendothelial cell; M, smooth muscle cell; MP, macrophage; N, nerve; P, pigment cell; PC, pericyte; V, vein; arrows, unspecialized nerve endings.

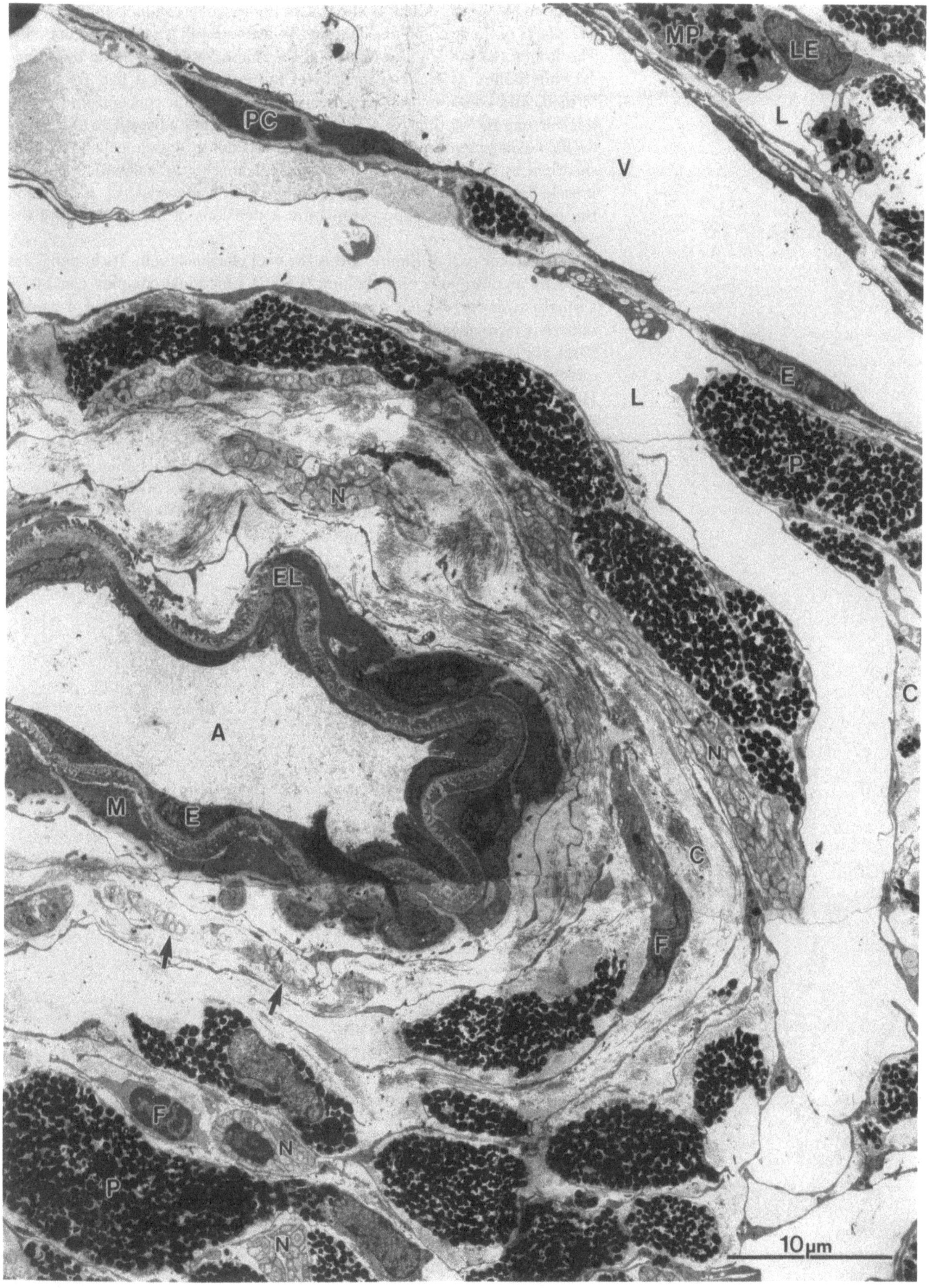
MP
LE
PC
L
V
L
E
P
N
EL
C
A
N
M
E
C
F
F
N
P
N
10µm

Figure 58 depicts details of the wall of the vascular channels in the *stroma vascularis* of the cynomolgus choroid. An arterial wall is seen in Figure 58A. The lumen (L) is at the lower part of the micrograph. It is lined by a typical endothelium (EC). The nucleus of the endothelium is flat, irregularly formed, and contains dark islands of heterochromatin concentrated at the nuclear margin. A thick (1.4 μm) elastic lamina lies adjacent to the endothelial basilar membrane. Three profiles of smooth muscle cells (MC), ensheathed by their own basilar membrane, belong to the incomplete circular muscle layer of this small artery. Some elastic material (E) is between the muscle cells (MC) and cytoplasmic extensions of fibrocytes (F) of the adventitia.

In the lower part of Figure 58B is the wall of a small vein. Its lumen (L) is lined by a continuous endothelium (EC). A junctional complex consists of a single stranded *zonula occludens* (tight junction) (arrow) and a *zonula adherens* (arrowhead). Pinocytotic vesicles (P) are seen at the luminal and basal surfaces of the endothelial cell. A pericyte is separated from the endothelium by a basilar membrane. This cell too has pinocytotic vesicles. Hemidesmosomes (H) attach the pericyte to its basilar lamina in the perimeter of the vein.

In the upper part of Figure 58B is the lumen (L) of a lymphatic cleft. Its endothelial lining (LE) is anchored by filaments (A) in the basilar membrane of the vein. Such anchoring filaments are characteristics of lymphendothelial cells (Leak and Burke 1968). Adjacent lymphendothelial cells are attached to each other by adhering junctions (arrowheads).

Figure 58. Cross sections through wall of choroidal vascular channels, cynomolgus monkey. (A) Arterial wall, parafoveal region, ×9,100. (B) Wall of vein and of lymphatic space, peripheral sample, ×49,000. A, anchoring filaments; B, basilar membrane; E, elastic material; EC, endothelial cell; F, fibrocyte; H, hemidesmosome; L, lumen; LE, lymphendothelium; M, mitochondrion; MC, smooth muscle cell; MP, junctional microplica; P, pinocytotic vesicle; PC, pericyte; RER, rough endoplasmic reticulum; arrow, *zonula occludens*; arrowheads, *zonula adherens*.

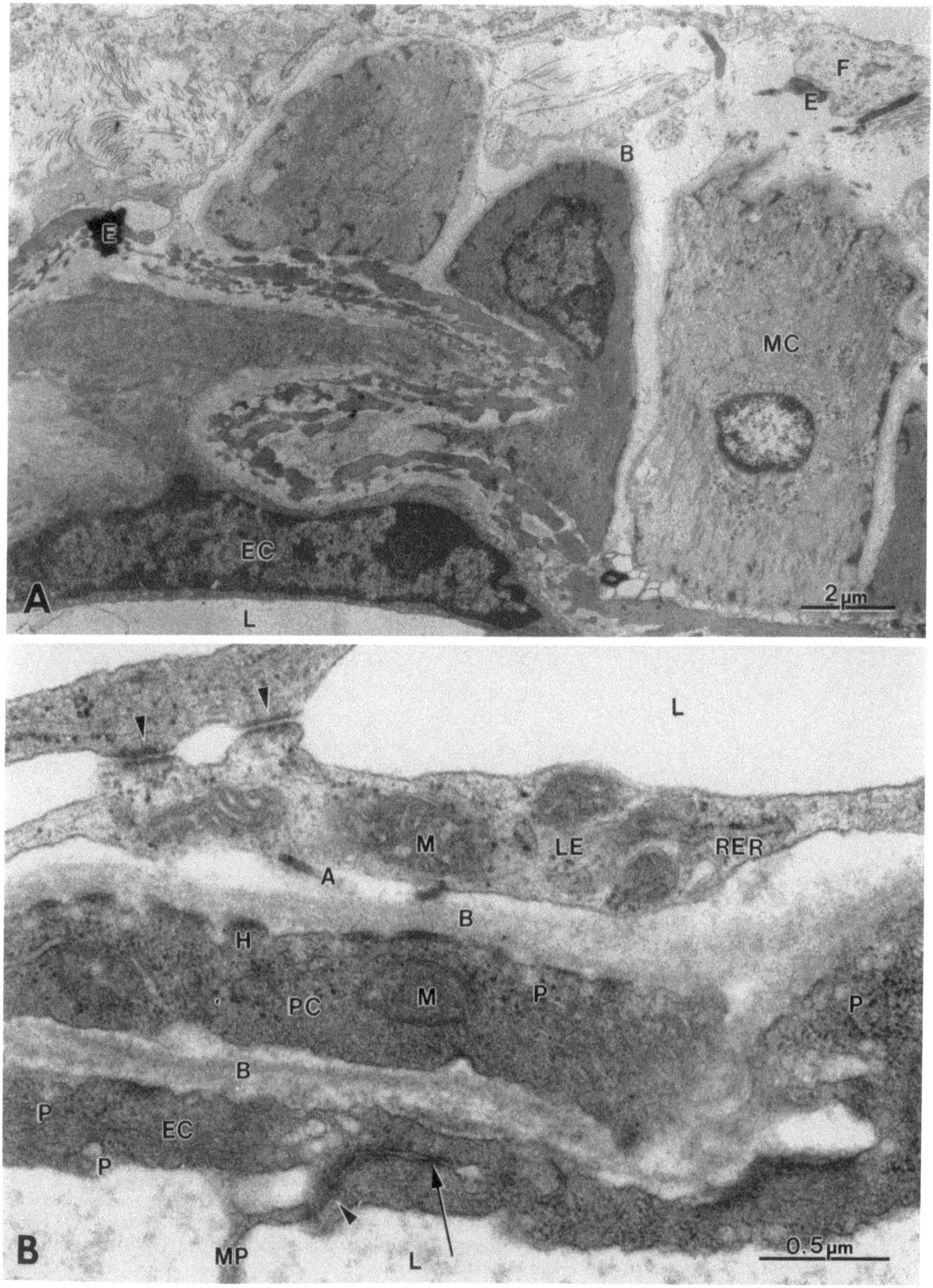
F
E
B
E
MC
EC
A
L
2 µm
L
M
LE
RER
A
B
H
PC
M
P
P
B
P
EC
P
B
MP
L
0.5 µm

Elements of the choroidal lymphatic endothelia are seen in **Figure 59**. In the central part of the *stroma vascularis*, flat sheaths of endothelial lamellae (LE) line multilayered empty clefts (Fig. 59A). The lamellae are connected to each other by junctional complexes that occasionally include tight junctions (Krebs and Krebs 1988b). They contain pinocytotic vesicles (P). A macrophage (M), containing lysosomes (LS) and phagocytized pigment granules, fills the lumen of a lymphatic cleft (Fig. 59A).

Elongated, electron-dense bodies (arrowhead), reminiscent of Weibel-Pallade bodies, are seen occasionally in the endothelial cytoplasm. The nature of these structures must remain unexplained. They may not be Weibel-Pallade bodies, because these organelles are believed to be diagnostic of endothelia in arteries and blood capillaries and have not been reported to occur in lymphendothelia. A review of their structure, function, and significance for ultrastructural pathology can be found in Ghadially (1982, pp. 623 ff.).

There are large openings in the endothelial walls (arrows). Such openings occur frequently in lymphatic endothelia (Leak 1976). In Figure 59B, an opening (arrow) is seen large enough for some collagen to enter the lumen (L). However, in general, collagen (C) is restricted to the interstitium outside the lymphatic spaces. Elastic material (E) and nerves (N) are seen in the neighborhood of the lymphendothelial cell (LE). The ubiquitous melanocytes of the monkey choroid (PC) often come close to the external surface of the lymphatic channels.

Figure 59. Lymphatic endothelium in choroid of foveal region. (A) Baboon, ×15,000. (B) Cynomolgus monkey, ×7,900. C, collagen; E, elastic material; J, junctional complex; L, lumen; LE, perikaryon of lymphendothelial cell; LS, lysosome; M, macrophage; N, nerve; P, pinocytotic vesicle; PC, pigment cell; arrow, openings in lymphendothelial wall; arrowhead, elongated, electron-dense body. (Section of Fig. 59 B was published in Krebs and Krebs (1988b), copyright 1988, American Medical Association.)

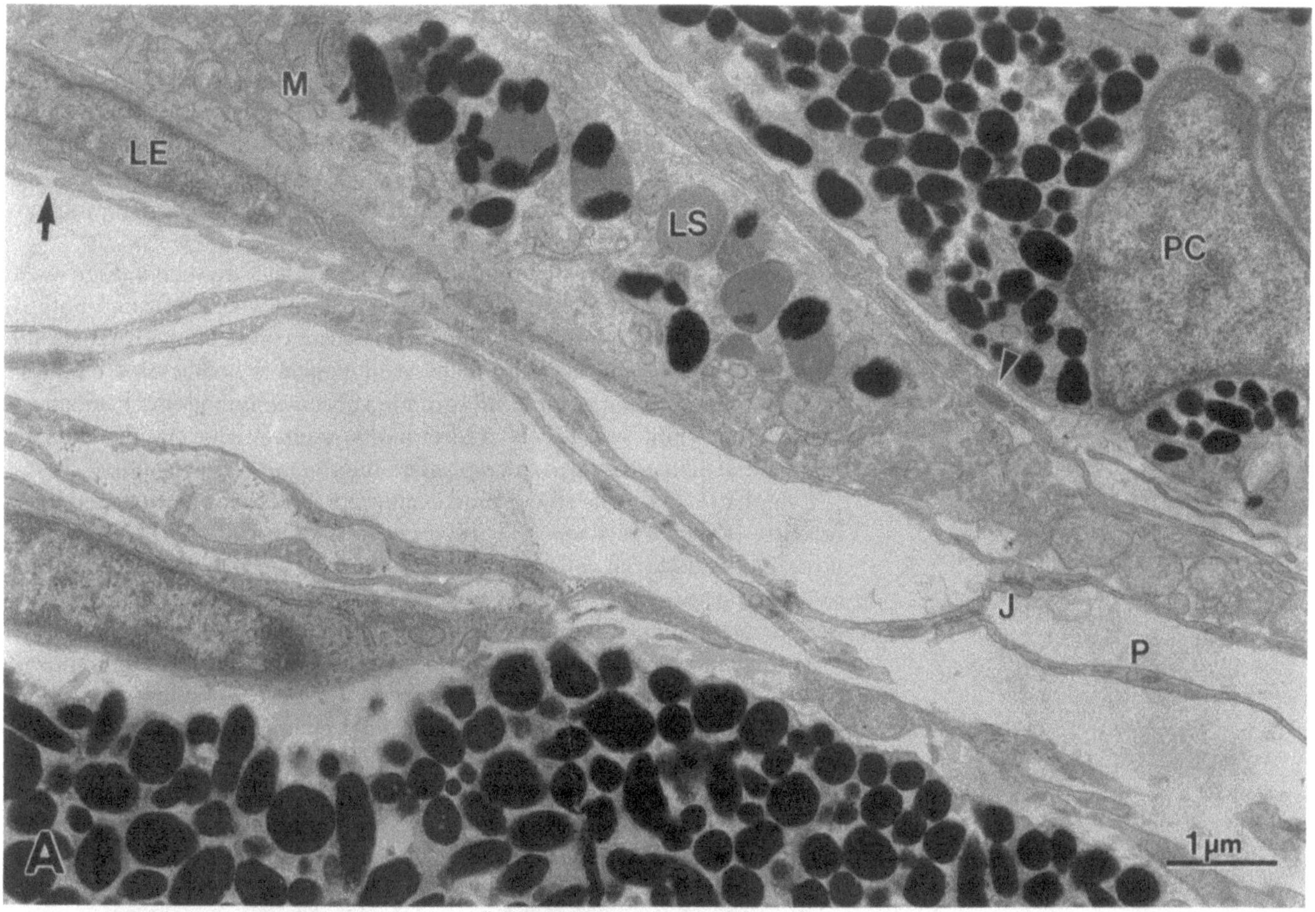
M
LE
LS
PC
J
P
A
1 µm

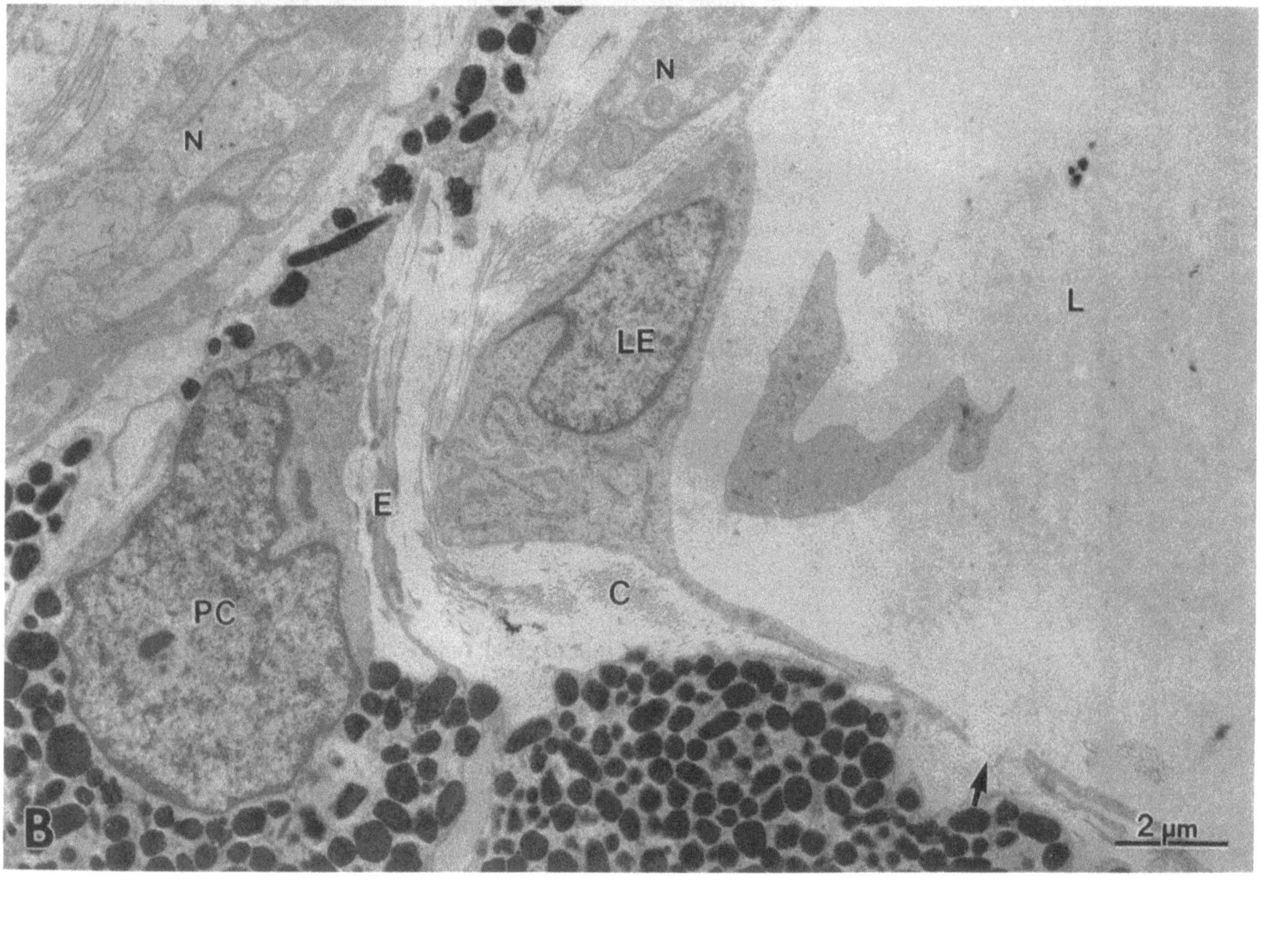
N
N
LE
L
E
PC
C
B
2 µm

Several sheaths of plate-like cells cover the choroid (**Fig. 60A**). These cells are attached to each other by adhering junctions (straight arrows). A few patches of basilar membrane-like material (BL) are seen, sometimes adjacent to a hemidesmosome-like density in the plasmalemma (arrowhead). Very little collagen (C) is present between the lamellae of the suprachoroid. The junctions between adjacent suprachoroidal cells can be quite elaborate (Fig. 60B). Besides adhering junctions (straight arrows), occluding junctions (curved arrow) occur.

Cytoplasmic features of lymphendothelial cells (LE) are seen in Figures 60 C through 60E. The nucleus (N) contains marginally concentrated heterochromatin. A few cisternae of rough endoplasmic reticulum (RER), some mitochondria (M), and the afore mentioned elongated bodies (B) are seen in their cytoplasm. Their cytoplasmic ground substance is moderately electron-dense. Pinocytotic vesicles (P) are frequently seen. Anchoring filaments (A) originate at densities in the cytoplasmic membrane (arrowhead).

Lymphatic spaces in the primate choroid have been recognized only recently (Krebs and Krebs 1988b). Feeney and Hogan (1961a), however, described endothelial cells in the human choroid; but they decided later not to interpret these cells as endothelial. Gomez et al. (1988) saw "veins with the characteristics of lymphatic vessels" in the choroid of the rabbit.

As the connections of the lymphatic spaces described here are not known, their function can only be speculated upon. However, some functions of the choroid would be facilitated by a viable fluid channeling system. It is conceivable that the lymphatic spaces of the choroid are continuations of the trabecular meshwork in the anterior chamber angle. The similarity between the endothelial cells of both systems is striking in any event. The aqueous fluid of the eye is drained partly through interstitial routes into the suprachoroid (Bill 1985). According to DeLaMotte (1978) and to Gomez et al. (1988), also cerebrospinal fluid is absorbed in the choroid. The porous choroidocapillaris is certainly capable of absorbing these fluids, and the choroidal lymphatic spaces would provide the channels for them. Thus, the lymphatic system of the choroid may be restricted locally and possibly is not connected to the lymphatics of the general circulatory system.

However, molecules too large to permeate the pores in the choroidal endothelium would have to be removed from the interstitium by other means. This could be accomplished by transendothelial transport via vesicles (Casley-Smith 1983). The abundance of vesicles in the choroidocapillary endothelium suggests this transport route. Alternatively, lymph-bound substances must be transferred into the general lymphatic system. Structural connections between the interstitium of the choroid and the lymphatics of the rest of the body have not been found yet. But, it was reported repeatedly that markers injected into ocular tissue were recovered from the cervical lymph node (Bill 1985; DeLaMotte 1978). The possible routes of this transport have been discussed by Grüntzig (1983).

Figure 60. Cells of suprachoroid and endothelia of lymphatic spaces in baboon choroid. (A, B) Suprachoroid, radial section, peripheral sample. (A) ×37,500. (B) ×51,300. (C, E) Lymphendothelia. (C) Foveal sample, ×25,600. (D) Foveal sample, ×26,200. (E) Peripheral sample, ×45,600. A, anchoring filaments; B, elongated body; BL, basilar membrane; C, collagen; L, lumen; LE, lymphoendothelial cell; M, mitochondrion; N, nucleus; P, pinocytotic vesicle; RER, rough endoplasmic reticulum; straight arrows, adhering junction; curved arrow, occluding junction; arrowhead, density in plasmalemma.

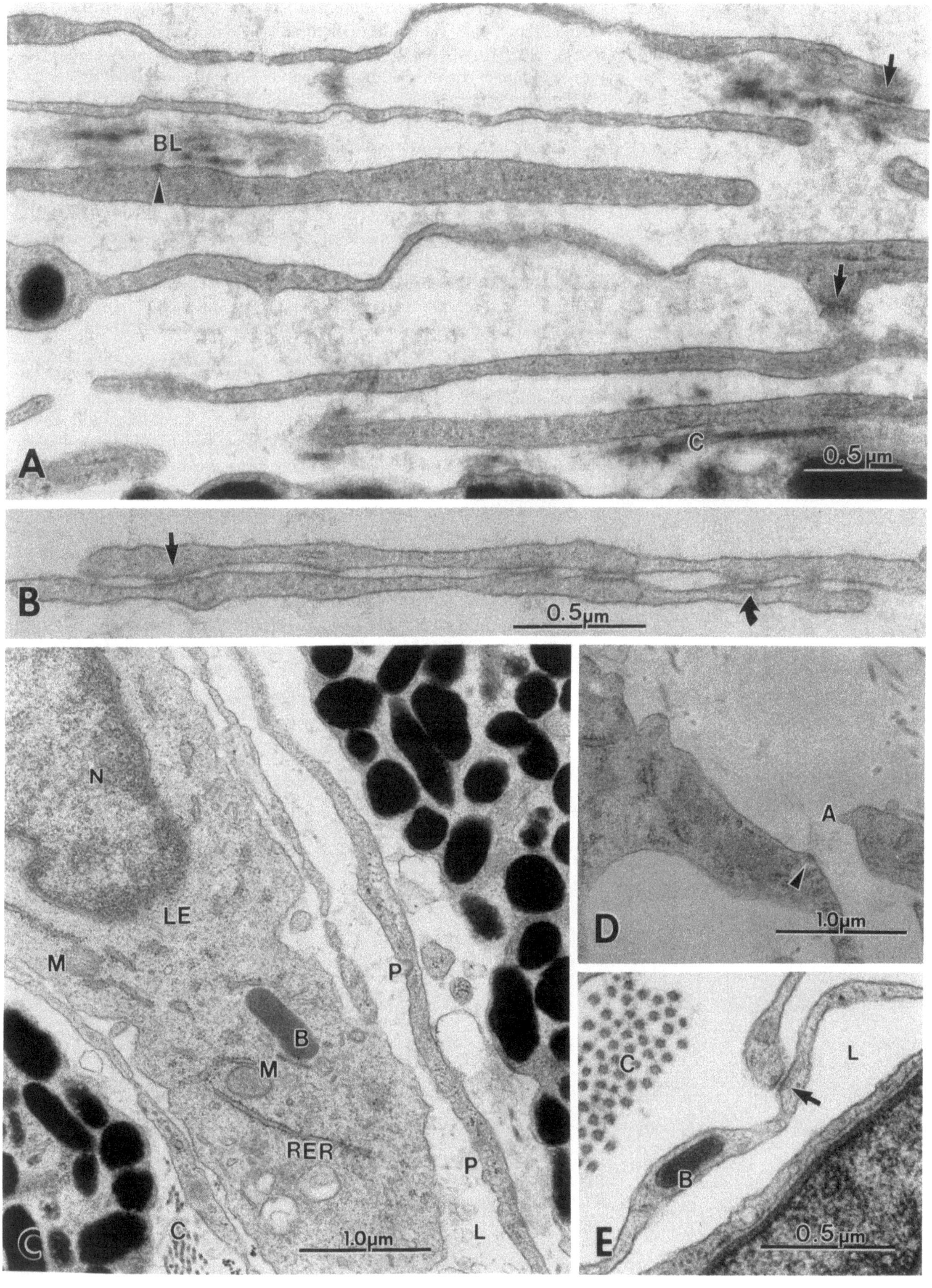
BL
C
0.5 µm
A
0.5 µm
B
N
LE
M
P
B
M
RER
P
L
C
1.0 µm
C
A
1.0 µm
D
C
L
B
0.5 µm
E

Choroidal Nerves

The presence of nerves in the choroid has long been known (Kurus 1955, Wolter 1960), and their ultrastructure has been described in detail by Feeney and Hogan (1961b). The high spatial density of these nerves (see Fig. 48) renders the choroid unique among primate connective tissues.

There are sympathetic and parasympathetic fibers (Ruskell 1971) in the choroid. Among other vasoactive polypeptides, substance P and the ubiquitous vasoactive intestinal polypeptide (VIP) have been demonstrated in the choroid (Bill 1985). No sensory innervation could be found in the primate choroid by Bergmanson (1977).

The nerves are restricted to the *stroma vascularis* of the choroid (Torczynski 1982). Early reports of nerve fibers in the choroidocapillaris (Wolter 1960) could not be confirmed. Such findings can be explained by the presence of the thin collagen fibers in the fiber system of the choroidocapillaris (see Fig. 49). These fibers might be argyrophilic and may have reacted with the silver stains used to demonstrate nerve fibers by light microscopy.

Figure 61 depicts the exterior part of the choroid underneath the suprachoroid (SU). The electron micrograph is from a radial section through the foveal region of the baboon. A large nerve (N_1) is cut longitudinally. It contains a large myelinated axon (MY), filled with neurotubules and a few mitochondria (M). The myelin producing Schwann cell (SC) is separated from Schwann cells of nonmyelinated axons by a distinct gap. The nonmyelinated axons are embedded in an electron-dense matrix of Schwann cell cytoplasm (asterisk). Mitochondria and neurotubules (NT) are seen in the axonal cytoplasm. The peripheral axons in the nerve designated as N_1 occasionally contain small dark vesicles (arrow). A perineurion (PN) of several layers of sheath-like cells envelops the larger nerves in the choroid.

The smaller nerves (N_2–N_5) lack a perineurion. The basilar membrane of the Schwann cell is in direct contact with the surrounding tissue. Between the nerves are choroidal melanocytes (P). A perikaryon of a suprachoroidal cell (SP) is seen in the lower right of Figure 61.

Figure 61. Radial section through exterior choroid in foveal sample of baboon, ×7,100. M, mitochondrion; MY, myelin; N_1–N_5, nerves; NT, neurotubuli; P, pigment cell; PN, perineurium; SC, Schwann cell; SP, perikaryon of suprachoroidal cell; SU, suprachoroid; arrow, swelling of axon containing electron-dense vesicles; asterisk, Schwann cell cytoplasm.

SC
M
NT
MY
NT
M
PN
P
N5
SU
SP
PN
N4
N1
N2
N3
5 µm

Two nerves of the central region of the *stroma vascularis* of the baboon choroid are depicted in **Figure 62**. The larger nerve (Fig. 62A) is ensheathed by a six to eight layered perineurium (PN). A large myelinated axon (MY) with its Schwann cell is isolated from the other fibers by a gap containing collagen and the basilar membranes of Schwann cells. The diameter of this axon is 1.1 μm, and its myelin sheath is 0.4 μm thick.

The nonmyelinated fibers in the nerve shown in Figure 62A vary in diameter from 0.2 to 1.2 μm. They are embedded in the cytoplasm of Schwann cells (asterisk). The plasmalemma invaginates into a mesaxon that often is accentuated by an adhering junction (arrowhead). A basilar lamina (straight arrows) covers the Schwann cells completely. The interstices between the Schwann cell domains contain longitudinally oriented collagen fibers. Collagen bundles are also seen between the lamellae of the perineurium.

Figure 62B shows a small nerve. Note that Fig. 62B is at a higher magnification than Fig. 62A. The nuclear region (N) of the Schwann cell is in the plane of the section. The few interstices between the Schwann cell islands contain only little collagen (C) and a basilar membrane (straight arrow). The axons are not always covered completely by Schwann cell cytoplasm (curved arrows). The smallest axons in Figure 62B are 0.1 μm in diameter, and the largest, 1.3 μm.

Quantitative analysis of radial sections of the foveal region of the baboon choroid revealed that the larger nerves with perineurium contained 20 to 80 axons, one of which was usually myelinated. The smaller nerves, without a perineurium, contained 4 to 36 axons; 1,800 nerve profiles were counted per square millimeter. The average distance between nearest neighbors was 19 μm, the values ranging from 4.6 to 39 μm.

Figure 62. Cross sections of choroidal nerves, foveal sample, baboon. (A) ×10,700. (B) ×16,800. C, collagen; M, mitochondrion; MY, myelin; N, nucleus of Schwann cell; PN, perineurium; SER, smooth endoplasmic reticulum; SC, Schwann cell; straight arrows, basilar membrane; curved arrows, axon not fully enveloped by Schwann cell cytoplasm; arrowhead, mesaxon with small adhering junction; asterisk, Schwann cell cytoplasm.

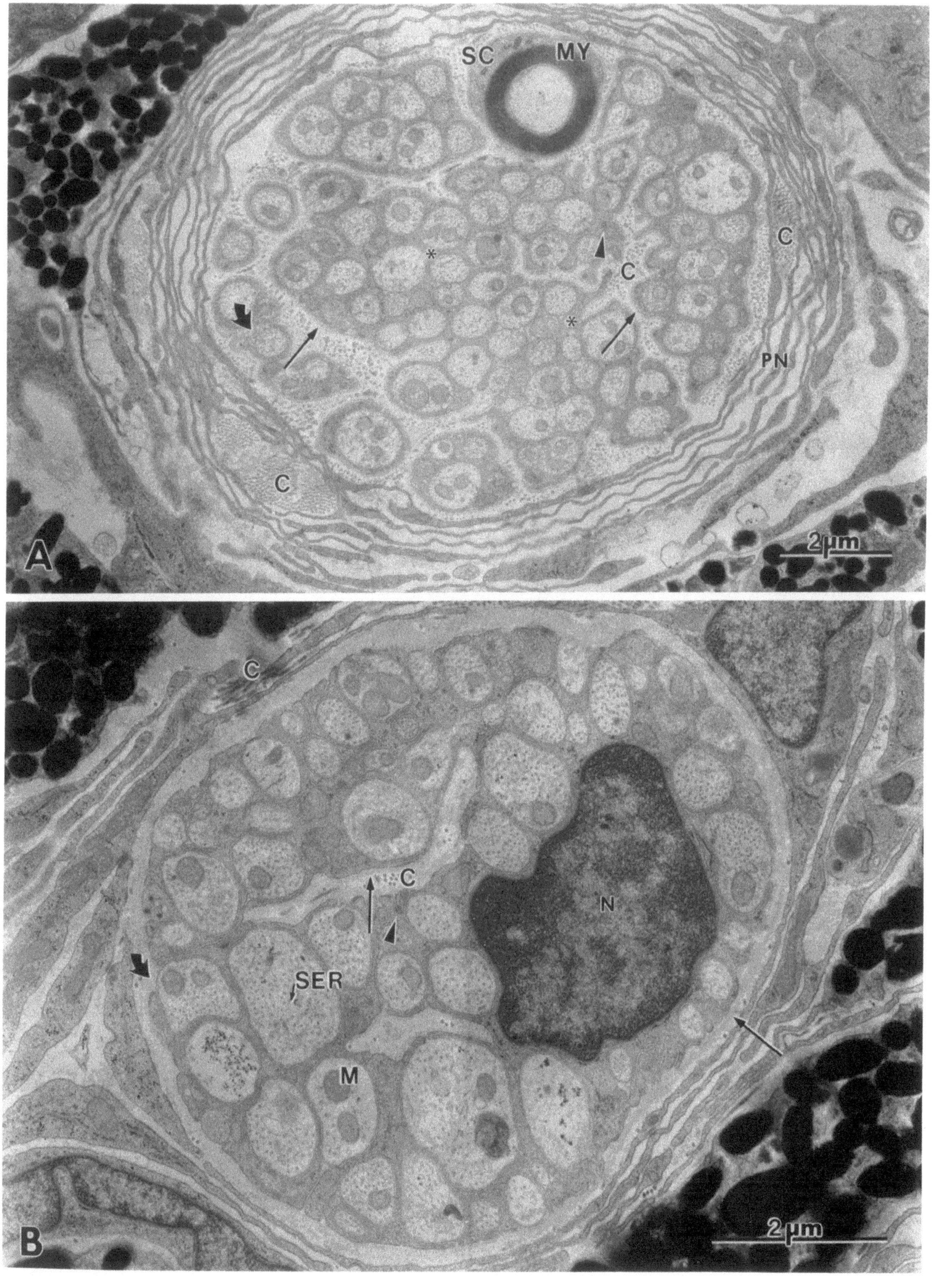
SC
MY
C
C
PN
C
2μm
A
C
C
N
SER
M
2μm
B

Figure 63A shows a small nerve in the internal part of the *stroma vascularis*. It demonstrates the closest position of nerves to the choroidocapillaris, encountered in these specimens. The choroidocapillaris (CC) is seen in the lower left corner of the micrograph. Note that the tangential plane (thick arrows), containing the external leaflet of the fiber system, is oriented obliquely here. Fibrocytes (F) and elastic material (E) are seen adjacent to the fiber system. The lumen of a draining vein (V) is lined by an endothelium (EN) studded with pores (thin arrows). The significance of endothelial pores in larger blood vessels close to the choroidocapillaris is discussed in the text discussing Figure 55.

A nerve (N) runs diagonally through Figure 63A. Its axons contain neurotubules and mitochondria (M). The electron-dense cytoplasm of the Schwann cells is seen between the axons. The wall of an artery (A) lies next to the nerve in the upper right of Figure 63A. The wall consists of endothelium (EN), smooth muscle cells (MC), and an external elastic layer (E). The artery very likely is the target of this small nerve.

Figure 63B depicts a few nerve endings adjacent to a small vein (V) in the central part of the peripheral cynomolgus choroid. Nerve endings at veins are considerably rarer than those at arteries. One of these endings contains solidly electron-dense vesicles (D). These vesicles may contain noradrenalin. The middle ending displays many clear (C) and a few dense core (DC) vesicles, suggesting that it belongs to a peptigergic neuron (Hökfelt et al. 1980). Some glycogen (G) is seen in the nerve terminals. Schwann cell cytoplasm (S) is seen between the nerve endings. However, the largest part of their surface is exposed to the extracellular space. A basilar membrane loosely surrounds the group of terminals. The basilar membrane of the venule is multilayered.

The functional role of the choroidal nerves is not completely understood. What is known about their function was reviewed by Bill (1985). The sympathetic system in the choroid acts vasoconstrictively. It apparently protects the choroid from being affected by elevation in the systemic blood pressure. The antagonistic vasodilatation probably is mediated by the vasoactive intestinal polypeptide (VIP).

These findings and similar effects of nervous stimulation (Bill 1985) cannot sufficiently explain the high spatial density of nerves in the choroid. A hypothetical explanation was offered by Krebs and Krebs (1990). In response to pressure fluctuation in the blood as well as in the vitreous, the nerves may regulate the vascular tone of choroidal blood vessels to create an outward pull on the fiber system of the choroidocapillaris. This pull stabilizes an equilibrium between inward and outward forces at a level that ensures optimal perfusion of the capillary sinus. Be that as it may, more investigations are urgently needed to shed some light on this important aspect of ocular function.

Another enigma of the choroid is the presence of large ganglion cells. Although they have not been seen in the material presented here, their

Figure 63. Nerve close to choroidocapillaris and nerve terminals at venule. (A) Nerve, venule, and artery, foveal region, baboon, ×5,500. (B) Nerve terminals, peripheral sample, cynomolgus monkey, ×36,300. A, artery; B, basilar membrane; C, clear vesicle; CC, choroidocapillary lumen; D, dense vesicle; DC, dense core vesicle; E, elastic material; EN, endothelium; F, fibrocyte; G, glycogen; M, mitochondrion; MC, smooth muscle cell; N, nerve; S, Schwann cell; V, vein; thin arrows, endothelial pores; thick arrows, orientation of tangential plane containing outer leaflet of fiber system of choroidocapillaris.

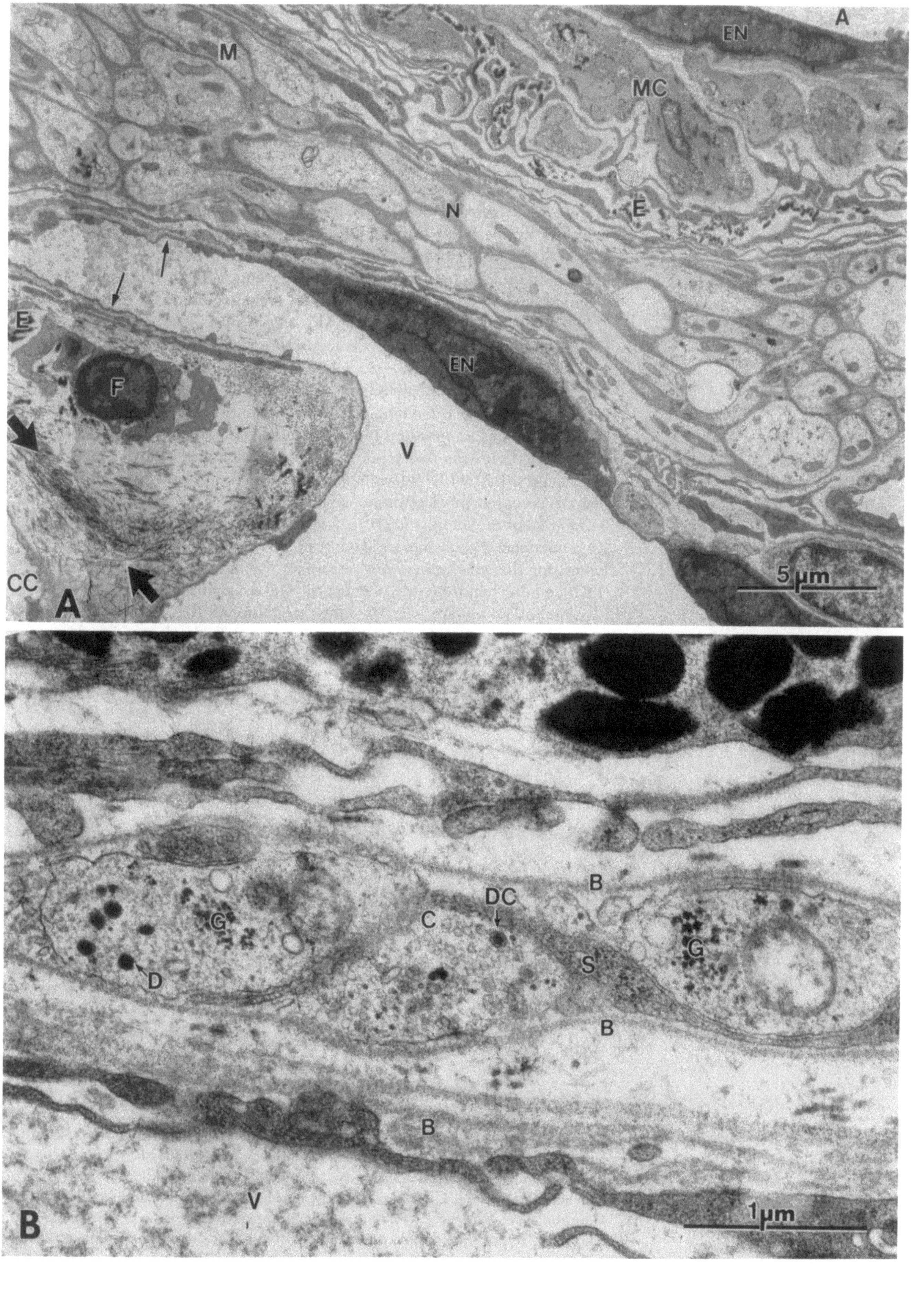
A
EN
M
MC
E
N
E
F
EN
V
CC
A
5 µm
B
DC
C
G
G
S
D
B
B
V
B
1µm

existence was documented by Hogan et al. (1972, p. 382 ff). They possibly occur anteriorly of the region that is designated "peripheral" here (Kurus 1955). The choroidal ganglion cells are described as large multipolar cells. Nothing is known about their connections.

Cells of Choroidal Connective Tissue

Feeney and Hogan (1961a) studied the choroidal cells by electron microscopy and found all the cells typical of connective tissue. A few examples of some of these cells are presented here.

Figure 64 shows an assortment of cells in the internal part of the choroid in the foveal region of the baboon. The level of the section is external from, but close to, the outer leaflet of the fiber system of the choroidocapillaris. Two of the plate-like fibrocytes (F) are seen in the upper right. They have an electron-dense cytoplasm containing rough endoplasmic reticulum, mitochondria, and Golgi fields. The fibrocytes are surrounded by collagen (C).

In the left-upper part of Figure 64 are a nerve (N) and part of a lymphocyte (LC). The nerve contains unmyelinated axons and Schwann cell cytoplasm. It is surrounded by a basilar membrane. A lymphendothelial cell (LY) is part of a series of lymphatic clefts (CL). The cell contains a few stacks of rough endoplasmic reticulum (RER), mitochondria (M), a lysosome that possibly contains an ingested pigment granule (L), and the cross section of an elongated electron-dense body (B). Some patches of basilar membrane-like material (arrow) lie adjacent to regions of the cytoplasmic membrane that are accented by electron-dense material.

Into one of the lymphatic clefts, a neutrophilic granulocyte (NG) extends its pseudopod (PS), which is devoid of cellular organelles. The granules typical for this cell type, a few cisternae of rough endoplasmic reticulum (RER), a mitochondrion (M), and a perikaryal concentration of smooth endoplasmic reticulum are seen in the cytoplasm of the granulocyte. The perikaryon of a pericyte (P) in the wall of a small blood vessel and some elastic material (E) are seen in the lower part of the electron micrograph.

Figure 64. Choroidal connective tissue cells, tangential section at level close to outer leaflet of fiber system, baboon, foveal region, ×13,800; B, cross section through elongated electron dense body; C, collagen; CL, lymphatic cleft; E, elastic material; F, fibrocyte; L, lysosome; LC, lymphocyte; LY, lymphendothelial cell; M, mitochondrion; N, nerve; NG, neutrophilic granulocyte; P, pericyte; PS, pseudopod; RER, rough endoplasmic reticulum; arrow, basilar membrane-like material.

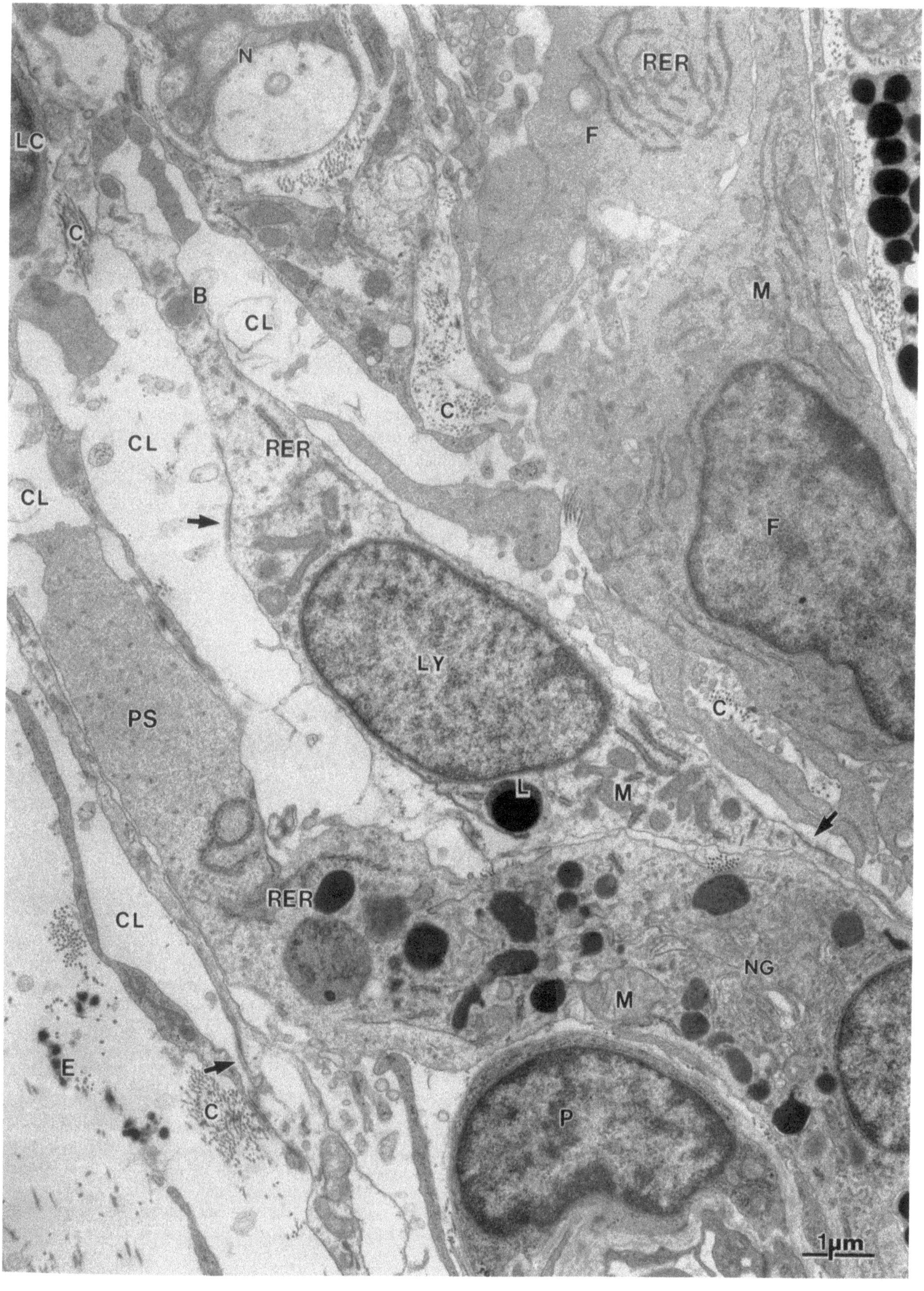
N
RER
LC
F
C
M
B
CL
C
CL
RER
CL
F
LY
PS
C
L
M
RER
CL
NG
M
E
C
P
1µm

Figure 65 depicts various cells. A fibrocyte of the outer leaflet of the choroidocapillary fiber layer is seen in Figure 65A. The plate-like cell is sectioned parallely to its flat surface. Several lysosomes (L) are in its cytoplasm. Its nucleus is indented, and the chromatin is concentrated marginally. Collagen fibrils (CO) are seen to originate from this cell.

Figure 65B is a melanocyte of the human choroid. It is surrounded by collagen fibrils. The melanin granules are globular and take up almost the entire space in the cytoplasm. Between them, mitochondria and some endoplasmic reticulum can be discerned. In a pigment-free zone adjacent to the nucleus, a lysosome (L) is visible. The chromatin in the globular nucleus is mostly marginally located.

Figure 65C shows a macrophage. Melanosomes of all shapes have been phagocytized; they fill composite lysosomes. Several lysosomes not containing melanosomes (L) lie between them. The elongated nucleus contains a large distinctive nucleolus. A plasmacell (Fig. 65D) displays the typical arrangement of its rough endoplasmic reticulum (RER). A mitochondrion (M) and the centriole (C) are among the usual organelles seen in the cytoplasm of this cell.

Eosinophilic granulocytes are seen occasionally in the choroid. The cell shown in Figure 65E contains the characteristic granules with crystalloid inclusions. Two profiles of the irregular nucleus are sectioned. A lymphocyte (LY; Fig. 65F), enclosed in a small draining venule, extends a pseudopod that penetrates into an invagination of an endothelial cell (E).

Figure 65. Cells in choroid. (A) Fibrocyte, human, peripheral sample, ×5,300. (B) Melanocyte, human, peripheral region, ×6,600. (C) Macrophage, baboon, foveal region, ×6,600. (D) Plasmacell, baboon, peripheral region, 7,000. (E) Eosinophilic granulocyte, baboon, parafoveal region, ×8,000; (F) Lymphocyte, baboon, foveal region, ×9,900. C, centriole; CO, collagen; E, endothelium; L, lysosomes; LU, lumen; LY, lymphocyte; M, mitochondrion; MP, macrophage; N, nucleus; NE, nucleolus; P, pigment cell; RER, rough endoplasmic reticulum; arrow, junctional complex.

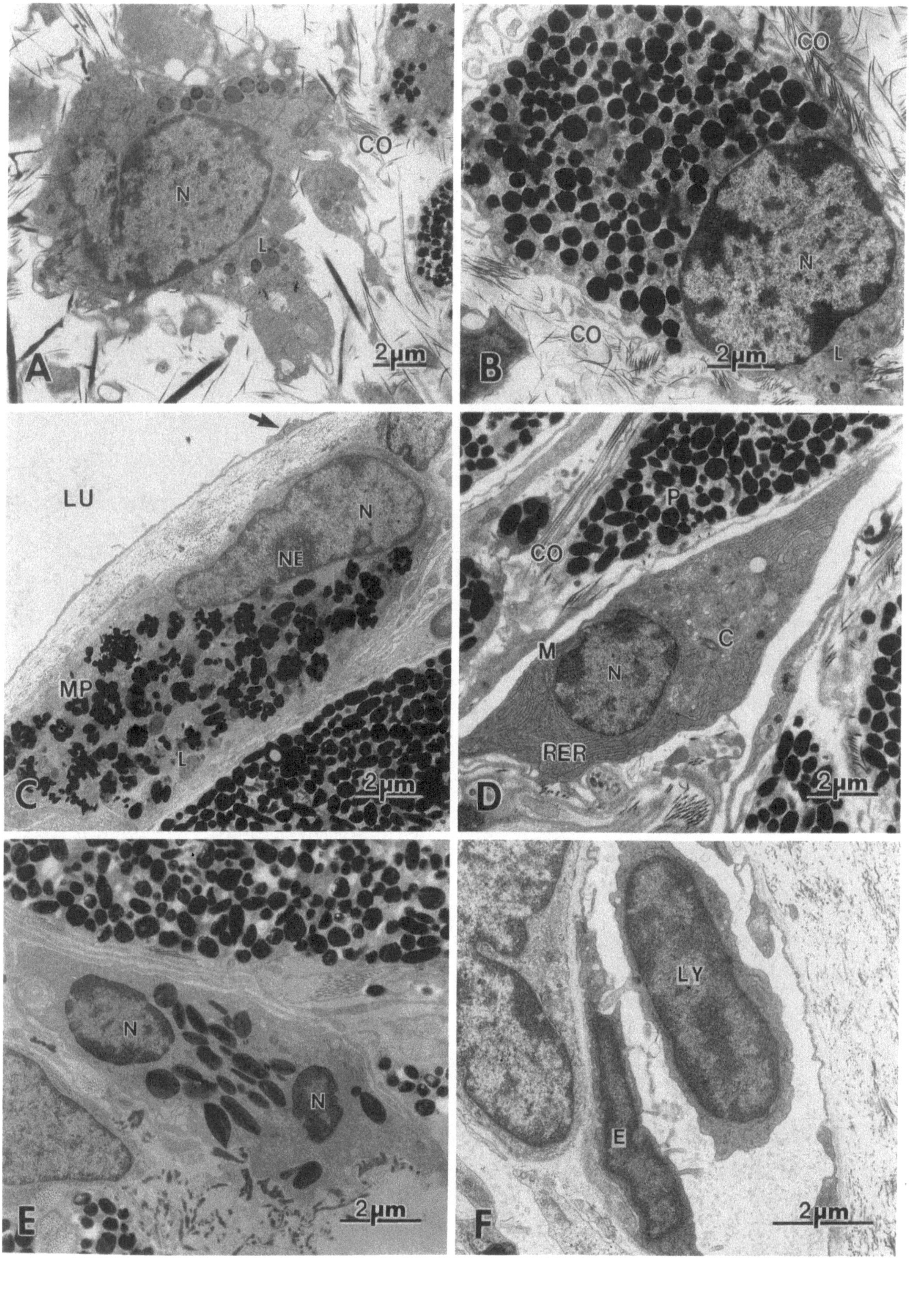
A
N
L
CO
2µm
B
N
L
CO
CO
2µm
C
LU
N
NE
MP
L
2µm
D
P
CO
C
M
N
RER
2µm
E
N
N
2µm
F
LY
E
2µm

References

Ahnelt PK, Kolb H, Pflug R (1987). Identification of a subtype of cone photoreceptor, likely to be blue sensitive, in the human retina. J Comp Neurol 255:18–34.

Barnstable CJ (1987). A molecular view of vertebrate retinal development. Molec Neurobiol 1:9–46.

Bergmanson JPG (1977). The ophthalmic innervation of the uvea in monkeys. Exp Eye Res 24:225240.

Bill A (1985). Some aspects of ocular circulation: Friedenwald lecture. Invest Ophthalmol Visual Sci. 26:410424.

Blanks JC, Hageman GS, Johnson LV, Spee C (1988). Ultrastructural visualization of primate cone photoreceptor matrix sheaths. J Comp Neurol 270:288–300.

Bok D (1985). Retinal photoreceptor-pigment epithelium interactions: Friedenwald lecture. Invest Ophthalmol Visual Sci 26:1659–1694.

Bone RA, Landrum JT, Fernandez L, Tarsis SL (1988). Analysis of the macular pigment by HPLC: Retinal distribution and age study. Invest Ophthalmol Visual Sci 29:843–849.

Borwein B (1981). The retinal receptor: A description. In: Enoch JM, Tobey FL (eds.), *Vertebrate Photoreceptor Optics*, pp. 11–81. Berlin: Springer-Verlag, (Springer Series in Optical Sciences).

Bunt AH (1971). Enzymatic digestion of synaptic ribbons in amphibian retinal photoreceptors. Brain Res 25:571–577.

Burnside B, Laties AM (1975). Pigment movement and cellular contractility in the retinal pigment epithelium. In: Zinn KM, Marmor FM (eds.), *The Retinal Pigment Epithelium*, p. 175. Cambridge: Harvard University Press.

Cajal RS (1904). Sistema visual – Organo receptor de la excitation visual ó retina. In: *Textura del systema nerviosa del hombre y de los vertebrados*, Tomo II. Madrid.

Casley-Smith JR (1983). The structure and functioning of the blood vessels, interstitial tissues, and lymphatics. In: Földi M, Casley-Smith JR (eds.), *Lymphangiology*, pp. 535–556. Stuttgart: Schattauer.

Chader GJ (1985). Cyclic nucleotides of the retina and pigment epithelium: Metabolism, control mechanisms, and involvement in physiological processes. In: Morgan WW (ed.), *Retinal Transmitters and Modulators: Models for the Brain, Vol. 1*, pp. 1–62. Boca Raton: CRC Press.

Chaitin MH, Burnside B (1989). Actin filament polarity at the site of rod outer segment morphogenesis. Invest Ophthalmol Visual Sci 30:2461–2469.

Chaitin MH, Carlson RB, Samara GJ (1988). Immunogold localization of actin in developing photoreceptor cilia of normal and rds mutant mice. Exp Eye Res 47:437–446.

Cohen AI (1961). The fine structure of the extrafoveal photoreceptors of the rhesus monkey. Exp Eye Res 1:128–136.

Curcio CA, Sloan KR, Packer O, Hendrickson AE, Kalina RE (1987). Distribution of cones in human and monkey retina: Individual variability and radial assymmetry. Science 236:579–580.

Das SR, Bhardwaj N, Gouras P (1990). Synthesis of retinoids by human retinal epithelium and transfer to rod outer segments. Biochem J 268:201–206.

Daw NW, Jensen RJ, Brunken WJ (1990). Rod pathways in mammalian retina. Trend Neurosci 13:110–115.

Dawson WW, Stratton RD, Hope GM, Parmer R, Engel HM, Kessler MJ (1986). Tissue responses of the monkey retina: Tuning and dependence on inner layer integrity. Invest Ophthalmol Visual Sci 27:734–745.

DeLaMotte DJ (1978). Removal of horseradish peroxidase and fluorescein-labelled dextran from CSF spaces of rabbit and optic nerve. A light and electron microscope study. Exp Eye Res 27:585–594.

Dowling JE (1987). *The Retina: An Approachable Part of the Brain.* Cambridge: Harvard University Press.

Enoch JM (1961). Nature of transmission of energy in the retinal receptors. J. Opt. Soc. Am. 51:1122–1126.

Famiglietti EV, Kolb H (1975). A bistratified amacrine cell and synaptic circuitry in the inner plexiform layer of the retina. Brain Res 84:293–300.

Fawcett DW (1966). *The Cell: Its Organelles and Inclusions.* Philadelphia and London: Saunders.

Feeney L, Hogan MJ (1961a). Electron

microscopy of the human choroid: I. Cells and supporting structures. Am J Ophthalmol 51:1057–1072.
Feeney L, Hogan MJ (1961b). Electron microscopy of the human choroid: II. The choroidal nerves. Am J Ophthalmol 51:1072–1083.
Freddo TF, Townes-Anderson E, Raviola G (1980). Rod-shaped bodies and crystalloid inclusions in ocular vascular endothelia of adult and developing *Macaca mulatta*. Anat Embryol 158:121–131.
Gallego A (1986). Advances in horizontal cell terminology since Cajal. In: Gallego A, Gouras P (eds.), *Neurocircuitry of the Retina: A Cajal Memorial*, pp. 122–139. New York: Elsevier.
Geiger B (1983). Membrane-cytoskeleton interaction. Biochim Biophys Acta 737:305–341.
Ghadially FN (1982). Ultrastructural Pathology of the Cell and Matrix, London: Butterworths.
Gomez DG, Manzo RP, Fenstermacher JD, Potts DG (1988). Cerebrospinal fluid absorption in the rabbit. Graefe's Arch Clin Exp Ophthalmol 226:17.
Gouras P (1991). Precortical physiology of colour vision. In: Cronly-Dillon JR (ed.), *Vision and Visual Dysfunction*, Vol. 6, pp. 163–176. New York: MacMillan.
Grüntzig J (1983). The eye and the lymphatic system. In: Földi M, Casley-Smith JR (eds.), *Lymphangiology*, pp. 535–556. Stuttgart: Schattauer.
Hagins WA, Penn RP, Yoshikami S (1970). Dark current and photocurrent in retinal rods. Biophys J 10: 380–412.
Haley JE, Flood MT, Gouras P, Kjeldbye HM (1983). Proteins from human pigment epithelial cells: Evidence that a major protein is actin. Invest Ophthalmol Visual Sci 24:803–811.
Heimann K (1972). The development of the choroid in man: Choroidal vascular system. Ophthalmol Res 3:257–273.
Hogan JM, Feeney L (1961). Electron microscopy of the human choroid: III. The blood vessels. Am J Ophthalmol 51:184–197.
Hogan JM, Alvarado JM, Weddell JE (1972). *Histology of the human eye: An Atlas and Textbook*. Philadelphia: Saunders.
Hökfelt T, Johansson O, Ljundahl Å, Lunberg JM, Schultzberg M (1980). Peptidergic neurones. Nature (London) 284:515–521.
Kolb H, Lipetz LE (1991). The anatomical basis for colour vision in the vertebrate retina. In: Cronly-Dillon J (ed.), *Vision and Visual Dysfunction*, Vol. 6, pp. 128–145. New York: MacMillan.
Korn ED (1978). Biochemistry of actomyosin-dependent cell motility (a review). Proc Natl Acad Sci USA 75:588–599.
Korte GE, Burns MS, Bellhorn RW (1989). Epithelium-capillary interaction in the eye: The retinal pigment epithelium and the choriocapillaris. Int Rev Cytol 114:221–248.
Krebs IP, Krebs W (1989). Discontinuities of the external limiting membrane in the *fovea centralis* of the primate retina. Exp Eye Res 48:295–301.
Krebs W (1982). *Die Retina des Rindes: Ein Beitrag zur Kenntnis ihrer Feinstruktur und Untersuchungen an isolierten Stäbchenaußensegmenten*. Berlin: Parey (Fortschritte der Veterinärmedizin).
Krebs W (1985). The role of rod outer segment discs in vision. In: Duncker HR, Fleischer G (eds.), *Functional Morphology in Vertebrates*, pp. 711–713. Stuttgart: Fischer (Fortschritte der Zoologie, Vol. 30).
Krebs W, Krebs IP (1987). Quantitative morphology of the peripheral primate retina (*Macaca irus*). Am J Anat 179:198–208.
Krebs W, Krebs IP (1988). Ultrastructural evidence for lymphatic capillaries in the primate retina. Arch Ophthalmol 106:1615–1616 (Copyright 1988, American Medical Association).
Krebs W, Krebs IP (1989a). Quantitative morphology of the central fovea in the primate retina. Am J Anat 184: 225–236.
Krebs W, Krebs IP (1989b). Spindle shaped bodies in foveolar cones of the baboon retina. Proc Electron Microsc Soc Am 47:960–961.
Krebs W, Krebs IP (1990). The fiber system of the choroidocapillaris. Histol Histopathol 5:219–224.
Krebs W, Kühn H (1977). Structure of isolated bovine rod outer segment membranes. Exp Eye Res 25: 511–526.

Kühn H, Bennet N, Michel-Villaz M, Chabre M (1982). Interaction between photoexcited rhodopsin and GTP binding protein: Kinetic and stoichiometric analysis from light scattering changes. Proc Natl Acad Sci (USA) 78:6873–6877.

Kurus E (1955). Über ein Ganglienzellsystem der menschlichen Aderhaut. Klin Monatsbl. Augenheilkd 127:198–206.

Kuwabara T (1975). Species differences in the retinal pigment epithelium. In: Zinn KM, Marmor FM (eds.), *The Retinal Pigment Epithelium*, p. 58. Cambridge: Harvard University Press.

Leak LV (1976). The structure of lymphatic capillaries in lymph formation. Fed Proc 35:1863–1971.

Leak LV, Burke JF (1968). Ultrastructural studies on the lymphatic anchoring filaments. J Cell Biol 36:129–149.

Liebman PA, Entine G (1974). Lateral diffusion of visual pigment in photoreceptor disc membranes. Science 185:457–459.

Marmor MF, Lurie M (1979). Light-induced electrical responses of the pigment epithelium. In: Zinn KM, Marmor FM (eds.), *The Retinal Pigment Epithelium*, p. 226. Cambridge: Harvard University Press.

Masland RH (1988). Amacrine cells. Trends Neurosci 9:405–410.

Misotten L (1965). *The Ultrastructure of the Human Retina*. Bruxelles: Edition Arsia S.A.

Misotten L (1974). Estimation of the ratio of the cones and neurons in the fovea of the human retina. Invest Ophthalmol Visual Sci 13:1045–1049.

Nomina Anatomica (1983). Baltimore: Williams & Wilkins.

Nomura T (1976). Electron microscopic study of subsurface cisterns in the retinal ganglion cells. Jpn J Ophthalmol 20:186–192.

Ogden TE (1983). Nerve fiber layer of the primate retina: Thickness and glial content. Vision Res 23:581–587.

Ogden TE (1984). Nerve fiber layer of the primate retina: Morphometric analysis. Invest Ophthalmol Visual Sci 25:19–29.

Østerberg G (1935). Topography of the layer of rods and cones in the human retina. Acta Ophthalmol (Suppl) 6:1–102.

Owaribe K, Masuda H (1982). Isolation and characterization of circumferential microfilament bundles from retinal pigmented epithelial cells. J Cell Biol 95:310–315.

Ozanics V, Rayborn ME, Sagun D (1978). Observation on the fine structure of the developing primate choroid coat. Exp Eye Res 26:25–45.

Polyak SL (1941). *The Retina: The Anatomy and the Histology of the Retina in Man, Ape, and Monkey, Including the Consideration of Visual Functions, the History of Physiological Optics, and the Histological Laboratory Technique*. Chicago: University of Chicago Press.

Porter KR, Bonneville MA (1973). *Fine Structure of Cells and Tissues*. Philadelphia: Lea and Febiger.

Rager G (1979). The cellular origin of the b-wave in the electroretinogram – a developmental approach. J Comp Neurol 188:225–244.

Reichenbach A (1989). Glia: neuron index: Review and hypothesis to account for different values in various mammals. Glia 2:71–77.

Reichenbach A, Eberhardt W (1988). Cytographical specialization of enzymatically isolated rabbit retinal Müller (glial) cells: K^+ conductivity of the cell membrane. Glia 1:191–197.

Rhodin JAG (1974). *Histology*. New York: Oxford University Press.

Riepe RE, Norenburg MD (1977). Müller cell localisation of glutamine synthetase in rat retina. Nature (London) 268:654–655.

Ringvold A, Davanger M (1989). Lateral cell connections at the inner segment level of human parafoveal photoreceptors. Acta Ophthalmol 67:430–434.

Rochon-Duvigneaud A (1907). Recherches sur la fovea de la retine humaine et particulierement sur le bouquet des cones centreaux. Arch Anat Microsc 9:315–342.

Rodieck RW (1973). *The Vertebrate Retina*. San Francisco: Freeman.

Röhrenbeck J, Wässle H, Boycott BB (1990). Horizontal cells in the monkey retina: Immunocytochemical staining with antibodies against calcium binding proteins. Eur J Neurosci 1:407–420.

Ruskell GL (1971). Facial parasympathetic innervation of the choroidal blood-vessels in monkeys. Exp Eye Res 12:166–172.

Sameshima M, Uehara F, Ohba N (1987). Specialization of the interphotoreceptor matrices around cone and rod photoreceptor cells in the monkey retina, as revealed by lectin cytochemistry. Exp Eye Res 45:845–863.

Sandig M, Kalnins VI (1988). Subunits in *zonulae adherentes* and striations in the associated circumferential microfilament bundles in chicken retinal epithelial cells *in situ*. Exp Cell Res 175:1–14.

Schein SJ (1988). Anatomy of macaque fovea and spatial densities of neurons in foveal representation. J Comp Neurol 269:479–505.

Schultze M (1866). Zur Anatomie und Physiologie der Retina. Arch Mikrosk Anat 2:165–286.

Shanthaveerappa TR, Bourne GH (1965). Histological and histochemical studies of the choroid of the eye and its relations to the pia-arachnoid mater of the central nervous system and perineural epithelium of the peripheral nervous system. Acta Anat 61: 379–398.

Shapley R, Perry VH (1986). Cat and monkey retinal ganglion cells and their visual functional roles. Trends Neurosci 9:229–235.

Staehelin LA (1975). Structure and function of intercellular junctions. Int Rev Cytol 39:191–283.

Starck D (1975). *Embryologie*. Stuttgart: Thieme.

Steinberg RH (1987). Monitoring communications between photoreceptors and pigment epithelial cells: Effects of "mild" systemic hypoxia. Invest Ophthalmol Visual Sci 12:1888–1904.

Stone J (1981). *The Wholemount Handbook: A Guide to the Preparation and Analysis of Retinal Wholemounts*. Sydney: Maitland.

Stryer L (1988). *Biochemistry*, pp. 1027–1040. New York: Freeman.

Torczynski E (1982). Choroid and suprachoroid. in: Jakobiec FA (ed.), *Ocular Anatomy, Embryology, and Teratology*, pp. 553–585. Philadelphia: Harper and Row.

T'so MOM, Friedman E (1967). The retinal pigment epithelium: I. Comparative histology. Arch Ophthalmol 78:641–649.

Uhl R, Wagner R, Ryba N (1990). Watching G proteins at work. Trends Neurosci 13: 64–70.

Wässle H, Boycott BB, Röhrenbeck J (1990). Horizontal cells in the monkey retina: Cone connections and dendritic network. Eur J Neurosci 1:421–435.

Williams DR (1988). Topography of the foveal cone mosaic in the living human eye. Vision Res 28:433–454.

Wolter JR (1960). Nerves of the human choroid. Arch Ophthalmol 64: 120–124.

Zauberman H (1979). Adhesive forces between the retinal pigment epithelium and sensory retina. In: Zinn KM, Marmor MF (eds.), *The Retinal Pigment Epithelium*. Cambridge: Harvard University Press.

Zinn KM, Marmor MF (eds.), (1979). *The Retinal Pigment Epithelium*. Cambridge: Harvard University Press.

Zrenner E (1983). *Neurophysiological Aspects of Color Vision in Primates: Comparative Studies on Simian Retinal Ganglion Cells and the Human Visual System*. Berlin: Springer-Verlag.

Index

MIX
Papier aus verantwortungsvollen Quellen
Paper from responsible sources
FSC® C105338

If you have any concerns about our products,
you can contact us on
ProductSafety@springernature.com

In case Publisher is established outside the EU,
the EU authorized representative is:
Springer Nature Customer Service Center GmbH
Europaplatz 3, 69115 Heidelberg, Germany

Printed by Libri Plureos GmbH
in Hamburg, Germany